Klaus H. Sames (ed.)

Cryopreservation and Lifespan Extension

Human and Animal Projects and Results

This Volume contains contributions to the second scientific symposium of the German society of Applied Biostasis (DGAB) held in Dresden, Germany 2014

Klaus H. Sames (ed.)

CRYOPRESERVATION AND LIFESPAN EXTENSION
Human and Animal Projects and Results

ibidem-Verlag
Stuttgart

Bibliografische Information der Deutschen Nationalbibliothek
Die Deutsche Nationalbibliothek verzeichnet diese Publikation in der Deutschen Nationalbibliografie; detaillierte bibliografische Daten sind im Internet über http://dnb.d-nb.de abrufbar.

Bibliographic information published by the Deutsche Nationalbibliothek
Die Deutsche Nationalbibliothek lists this publication in the Deutsche Nationalbibliografie; detailed bibliographic data are available in the Internet at http://dnb.d-nb.de.

Cover picture: Cells of the CA1 band of rabbit hippocampus. KR8H washout solution. Vitrified; CPA removed by perfusion. Frames from a FIB-SEM stack of rabbit neuropil near the CA1 band of the hippocampus. Stack dimensions: 10µm × 10µm × 5µm. All processes appear well preserved and traceable throughout the stack. Reprinted with permission from Cryobiology Volume 71, Issue 3, December 2015, P. 448–458. https://doi.org/10.1016/j.cryobiol.2015.09.003. Open Access under Creative Commons (CC BY 4.0, s. https://creativecommons.org/licenses/by/4.0/)

∞
Gedruckt auf alterungsbeständigem, säurefreien Papier
Printed on acid-free paper

ISSN: 2195-5700

ISBN-13: 978-3-8382-0721-6

© *ibidem*-Verlag
Stuttgart 2018
Alle Rechte vorbehalten

Das Werk einschließlich aller seiner Teile ist urheberrechtlich geschützt. Jede Verwertung außerhalb der engen Grenzen des Urheberrechtsgesetzes ist ohne Zustimmung des Verlages unzulässig und strafbar. Dies gilt insbesondere für Vervielfältigungen, Übersetzungen, Mikroverfilmungen und elektronische Speicherformen sowie die Einspeicherung und Verarbeitung in elektronischen Systemen.

All rights reserved. No part of this publication may be reproduced, stored in or introduced into a retrieval system, or transmitted, in any form, or by any means (electronical, mechanical, photocopying, recording or otherwise) without the prior written permission of the publisher. Any person who does any unauthorized act in relation to this publication may be liable to criminal prosecution and civil claims for damages.

Printed in the EU

Preface

We urgently need scientific studies in cryonics obviously as long as we are unable to resuspend our patients. We need analyzes of our current methods and improvement in current procedures as well as development of new methods to overcome the many obstacles limiting human cryonics application today. One major of those obstacles is lack of acceptance of cryonics by the scientists and medics, before we provide evidence that, cryonics works. Demonstrating our scientific background and our project planning as well as discussion of biological problems and the outlooks to development in the near future may trigger cryonics progress.

We need analyses of organ structure on the micro-ahd nano scale. Main topics always will be the consequences of organ failure and resulting ischemia, which set the stage for damage of tissues and cells. These are crucial conditions at the start of human cryopreservation.

Main concerns during cooling and rewarming in connection with vitrification e.g. freezing, cracking devitrification and toxicity must be overcome by new methods.

Studies in the history of cryonics should provide insight in main obstacles of cryonics development

Since resuspension will only be able, if the stop of aging, repair of the body and rejuvenation will be reached, we must include areas like gerontology and nanotechnology.

There still is lack of independent cryonics research projects, which also becomes visible by lack of original papers in the current volume. Model organisms may proof to be of value for analysis of ageing and for antiageing methods.

Intravital repair may well be able one day to extend the life span up to an unlimited, even if not—as fare as foreseeable—to an eternal level.

June 2017, Senden at the Iller
Klaus H. Sames

Table of Contents

Preface 5

The Robert Ettinger Medal for Outstanding Merits
in the Field of Cryonics 9
    2014 in Dresden. Laudation by Peter Gouras to Saul Kent ... 9
    2010 in Goslar. Laudation by Ben Best to Robert Ettinger . 11

Robert L. McIntyre, Gregory M. Fahy
Aldehyde-Stabilized Cryopreservation 13

Pulver A., Artyuhov I.V, Artyukhov V.I., Tselikovsky A.V, Shamaev N.V., Pulver N.A, Peregudov A
Combined Approach to the Development of Protocol for Vitrification of Bulky Biological Objects 47

N. Shamaev, A. Pulver, D. Buslov
Uniform Heating of Multi-Structural Biological Objects by Means of Electric and Magnetic Fields' Phased Emitters . 57

Peter Gouras
Trying to Outwit Nature 61

Ramon Risco, Alberto Olmo and Pablo Barroso
New Advances in Organ Cryopreservation.
Electromagnetic Rewarming and Selective Targeting of Ice Nuclei 65

Ben Best
Forms of Cryopreservation Damage and Strategies for Prevention or Mitigation 75

Aschwin de Wolf
Identification, Validation, and Implementation of New Cryonics Technologies (an essay) 83

Dirk Nemitz
History of Cryonics—A Narrative Analysis of the Cryonics Magazin 95

Klaus Mathwig
Molecular Repair at Physiological Conditions? 105

Nadine Saul
Anti-Ageing and Pro-Longevity: What can We Learn from a Small Worm? A Methodical Overview 117

Klaus Sames
Definitions of Death 129

Contributors 153

The Robert Ettinger Medal for Outstanding Merits in the Field of Cryonics was awarded to its Second Recipient: Saul Kent

The laudation was held by Peter Gouras in Dresden, Oct 2014

Laudation

I have not been a close friend of Saul but I have known him and followed his career since 1964. I actually talked several times with his mother, a phantom person in the Cryonics movement. Saul Kent has been one of the most significant promoters of Cryonics in the world. He is a born and bred New Yorker who could not have been named more appropriately. Saul is one of the great evangelists of this movement. His epiphany came when he was a college student in 1964 while listening to Robert Ettinger on the radio and reading, *"The Prospect of Immortality"*. He became an instant cryonicist. The epiphany resembles his namesake, Saul of Tarsus, later called Saint Paul, who was blinded on the road to Damascus by a dazzling appearance of Christ. Saul of Tarsus recovered and went on to promote an immortality in heaven for good behavior. Our Saul not of Damascus but Brooklyn is less mystical, more scientific promising immortality right here on earth and doing it with a similar determination as Saint Paul.

Saul Kent's family name is just as appropriate. It's the family name of Clark Kent, alias Superman, the comic book hero of the 1940s. Superman could do impossible things that mere mortals could not. He could fly, lift tall buildings and was impossible to kill. Our Saul Kent may not be as s strong as Superman but may be more immortal.

One can ask whether Saul Kent knows what he's talking about. But if you converse with him, you will be quickly impressed with his intelligence and logic as he considers the strategy of Cryonics. He will argue that this is the best solution at the present time for immortality and has science on its side. After all embryos and small organs like rabbit kidneys can be deeply cooled and stored without compromising their function when re-warmed, why not brains or entire bodies?

But it is not conversation alone that reveals his intelligence. His actions demonstrate it even more. Saul Kent has been very successful in the competitive business world. He is the founder of a profitable company that promotes anti-aging medications, a growing field. Much of his business profits support scientific research in Cryonics. He supports 21st Century Medicine which is where our greatest research is being done by Greg Fahy and Brian Wowk, two of the most competent scientists in the field of Cryobiology.

But after almost 50 years of promoting Cryonics, Saul has become disappointed with its progress. He has written that it faces extinction because of its lack of growth. Saul, you should have never thought this was going to be easy. But there is growth. Compared to 1964 we have Cryonic organizations throughout the world; in Canada, England, Australia, Russia, France and even New Caledonia. We hope that awarding you the Ettinger medal reinvigorates your enthusiasm for Cryonics ...

Peter Gouras M.D.

The Medal had been awarded to its First Recipient Robert Ettinger

The laudation was held by Ben Best in Goslar 2010

To the laudation
In evidence-based cryonics (http://evidencebasedcryonics.org) Ben Best summarized it as follows

I accepted the medal on behalf of Mr. Ettinger, which meant that I had to make a speech.

    I said that Robert Ettinger is above all a man of ideas, who nonetheless also felt obliged to exert his influence in the physical world by, among other things, helping found the Cryonics Institute because he was not satisfied with what the other cryonics organizations were offering. I also said that Mr. Ettinger deserves a lot of credit for the creation of CI's fiberglass cryostats, something he is rarely credited for.

Ben Best

Aldehyde-Stabilized Cryopreservation

Robert L. McIntyre, Gregory M. Fahy[1]

Abstract

We describe here a new cryobiological and neurobiological technique, aldehyde-stabilized cryopreservation (ASC), which demonstrates the relevance and utility of advanced cryopreservation science for the neurobiological research community. ASC is a new brain-banking technique designed to facilitate neuroanatomic research such as connectomics research, and has the unique ability to combine stable long term ice-free sample storage with excellent anatomical resolution. To demonstrate the feasibility of ASC, we perfuse-fixed rabbit and pig brains with a glutaraldehyde-based fixative, then slowly perfused increasing concentrations of ethylene glycol over several hours in a manner similar to techniques used for whole organ cryopreservation. Once 65% w/v ethylene glycol was reached, we vitrified brains at −135 °C for indefinite long-term storage. Vitrified brains were rewarmed and the cryoprotectant removed either by perfusion or gradual diffusion from brain slices. We evaluated ASC-processed brains by electron microscopy of multiple regions across the whole brain and by Focused Ion Beam Milling and Scanning Electron Microscopy (FIB-SEM) imaging of selected brain volumes. Preservation was uniformly excellent: processes were easily traceable and synapses were crisp in both species. Aldehyde-stabilized cryopreservation has many advantages over other brain-banking techniques: chemicals are delivered via perfusion, which enables easy scaling to brains of any size; vitrification ensures that the ultrastructure of the brain will not degrade even over very long storage times;

---

[1] Reprinted with permission from Cryobiology Volume 71, Issue 3, December 2015, P. 448–458. https://doi.org/10.1016/j.cryobiol.2015.09.003. Open Access under Creative Commons (CC BY 4.0, s. https://creativecommons.org/licenses/by/4.0/)

and the cryoprotectant can be removed, yielding a perfusable aldehyde-preserved brain which is suitable for a wide variety of brain assays.

1. Introduction

The objective of the present studies was to demonstrate that cryobiological techniques can be used to enable brain banking to support the needs of fields such as the emerging domain of whole brain connectomics (Eisenstein 2009). Connectomics is the study of the totality of all neuronal connections in individual brains so as to better understand the functions of the brain and the emergence of mind (Seung 2012; Sporns 2014). Connectomics involves tracing the linkages between nerve cells; currently available connectomics methods often trace these linkages in fixed brains.

In contrast to the requirements of many cryopreservation protocols, brain banking for connectomics research does not necessarily need to preserve the biological viability of brain tissue; the primary criterion for success is instead to maintain the delicate ultrastructural appearance of the brain.

Our goals in developing a robust, general-purpose brain banking protocol for connectomics research were that such a procedure must:

1. Provide the highest quality of ultrastructural preservation with minimal distortion.
2. Easily scale to brains of any size.
3. Enable indefinite storage of whole brains, with no ultrastructural changes during storage.
4. Remain compatible with as many neuroanatomical assays as possible.

Currently, there is no brain banking technique that achieves all four of our goals: while there are many techniques that are used in practice to preserve brain tissue, they all fail to meet one or more of our four requirements.

The mainstay method for whole brain preservation involves perfusing the brain (Palay et al. 1962) with aldehydes (Griffiths 1993; Hayat 1991) and storing the fixed brain at a relatively warm temperature, 4 °C. While this technique has been used for decades, it does not guarantee static preservation of brain ultrastructure over the course of years, because the fixed brain is still chemically active at 4 °C and can undergo chemical and morphological degradation over long storage times (Hooshmand et al. 2014; Xie et al. 2011). Nor is it possible to simply store fixed brains at sub-zero temperatures to inhibit chemical degradation, because the resulting formation of ice will cause significant dehydration and mechanical damage to the ultrastructure of the brain (Fahy et al. 1984a). In fact, fixed tissue often suffers even more severe damage during freezing than unfixed tissue, although this can be somewhat mitigated by immersing fixed brain tissue in cryoprotectants (Rosene and Rhodes 1990).

Another method for whole brain preservation is the recently developed technique of whole brain staining and resin embedding (Mikula et al. 2012; Mikula, Denk 2015). This approach shows great promise for stabilizing brains in resin at room temperature indefinitely. However, current staining and embedding techniques are based on diffusion of highly viscous embedding materials inward from the brain surface over macroscopic distances through fairly dense fixed brain tissue. The time required to diffuse chemicals into a brain theoretically scales quadratically with brain mass, making whole brain staining and resin embedding impractical for larger brains (>10g). Even a hypothetical staining and resin embedding procedure which employed perfusion to avoid the quadratic slowdown would have to pre-commit to a particular sort of stain and resin, limiting its use as a general-purpose brain-banking protocol, because both staining and resin embedding are irreversible, and resin-embedded brains cannot be reperfused with other chemicals. For example, resin-embedded brains would not be compatible with techniques such as expansion microscopy (Chen et al. 2015) or the CLARITY protocol (Chung et al. 2013), which require perfusion of a hydrogel after aldehyde fixation.

A third method for brain preservation, vitrification, involves perfusing organs with very high concentrations of cryoprotective agents (CPAs) and then storing them at extremely low temperatures such as −135°C (Fahy et al. 1984a). The CPAs prevent ice formation during cooling and instead the entire system becomes a solid glass (Wowk 2010). Vitrification has the potential to preserve biological systems indefinitely with no ice-mediated damage while maintaining biological viability (Fahy, Wowk 2015), but cryoprotective additives (CPAs) can themselves be toxic (Clark et al. 1984; Fahy 2010) and cause osmotic dehydration if added too quickly.

Successful vitrification protocols must therefore make a compromise between minimizing exposure to toxic CPAs (by minimizing CPA equilibration times) and minimizing exposure to dehydration and osmotic stress (by maximizing CPA equilibration times) (Fahy et al. 2004). For cryoprotecting the brain, the problem of dehydration is particularly severe because of the blood–brain barrier (BBB), which limits the rate at which cryoprotectants can enter the brain, thus causing major osmotic brain shrinkage (Fahy et al. 1984b). For the purposes of connectomics, this dehydration is undesirable because it distorts the brain's ultrastructure and causes difficulties in tracing fine neural processes.

To address the limitations of the previous methods discussed, we conceived of a simple solution that meets our four brain banking goals: aldehyde-stabilized cryopreservation (ASC). We fixed brains using aldehyde perfusion, then gradually perfused those brains with sufficiently high concentrations of cryoprotectant to enable vitrification. The aldehydes immediately stabilize the fine structure of the brain to an extent sufficient for connectomics research, meeting our goal of high-quality preservation. Once the brain is fixed, cryoprotectant toxicity and other chemical insults are of minimal concern. Therefore we were able to add cryoprotectant more gradually and to include a surfactant to accelerate CPA introduction by breaking down the BBB, allowing us to achieve dehydration-free vitrification.

Biological time in vitrified systems is essentially arrested (Fahy et al. 1984a; Fahy and Rall 2007) enabling very long term storage. In principle, ASC is scalable to brains of any size because all chemicals can be delivered by perfusion, and the distance between capillaries is essentially independent of the size of perfused organs. Finally, cryoprotectants can be removed from the brain after warming to yield an aldehyde preserved brain, a common currency among connectomics assays: traditional resin embedding (Hayat 1991), the BROPA whole brain embedding protocol (Mikula and Denk 2015). CLARITY (Chung et al. 2013), the BrainBow method (Lichtmann et al. 2008), expansion microscopy (Chen et al. 2015), and immunocytochemical assays (Griffiths 1993) might all be compatible with aldehyde-stabilized cryopreserved brains.

In this paper, we describe our initial exploration of achieving good brain cryopreservation by the ASC method. We describe here the applicability of this method to both small (rabbit, 10g) and large (pig, 80g) brains. The results show exquisite preservation of anatomical detail in both models after vitrification and rewarming, with virtually no identifiable artifacts relative to controls.

2. Materials and methods

2.1. Animals

We used 37 adult (12–36 week old) male New Zealand White (NZW) rabbits (obtained from Charles River Laboratories) and 3 young (3–4 month old) female Yorkshire pigs (obtained from SNS Farms). The rabbits were used to refine the parameters of the ASC protocol as determined by electron microscopy of brain tissue, and the pigs were used to demonstrate preservation of larger brains using ASC. All procedures were approved by 21st Century Medicine's Institutional Animal Care and Use Committee and were in full compliance with USDA standards and guidelines for animal care.

2.2. Perfusion machine

Our rabbit cephalon perfusion machine consisted of two parts: 1) A mobile cart (washout cart) described in Fig. 1, which was present during the initial surgery and which we used to wash out blood and cool the rabbit brain, and 2) a computer-controlled perfusion apparatus (CPA circuit), which was entirely contained in a fume hood (Fig. 2), which we used for fixation and the cryoprotectant concentration ramp.

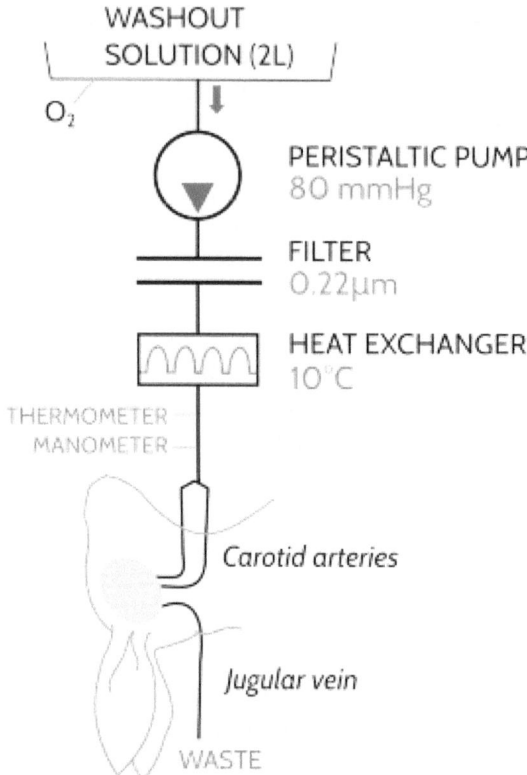

Fig. 1.
The washout cart was used to wash out the rabbit's blood, then to move the cephalon to the CPA circuit (Fig. 2) for fixation and cryoprotectant introduction. The washout circuit consisted of a 2L reservoir of oxygenated blood washout solution, a 0.22μm filter, heat exchanger (ice water and heat exchanger pump not pictured), and a thermometer and manometer to monitor perfusion parameters.

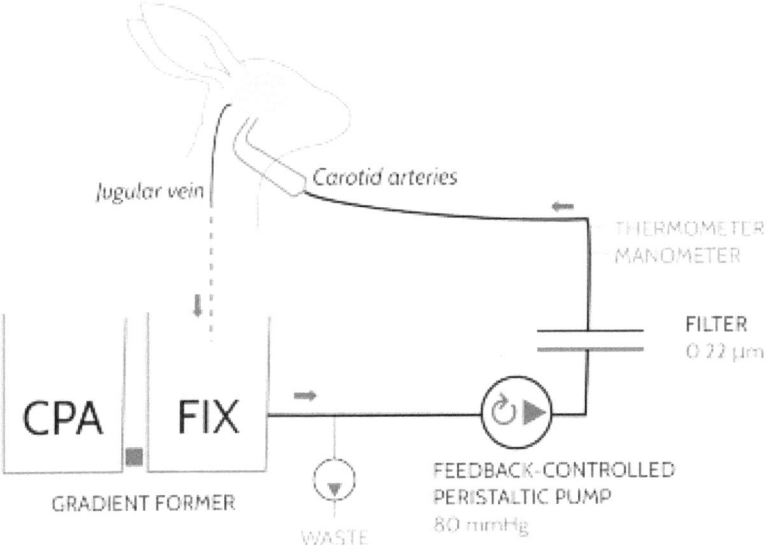

Fig. 2.
The CPA circuit was entirely contained in a fume hood, and was used to fix the rabbit brain and to introduce CPA. It consisted of a gradient generator, a computer-controlled peristaltic pump, a waste pump, a 0.22µm filter, and a digital manometer and thermometer. The gradient generator consisted of adjacent reservoirs of cryoprotectant solution (CPA) and fixative solution (FIX) connected by a short tube. The fixative reservoir was continuously stirred by a magnetic stir bar. A waste pump was used to lower the level of the gradient generator and create the gradient. The computer adjusted the flow rate of the feedback-controlled perfusion pump to maintain constant pressure of the cephalon. To mimic standard neurobiology procedures, fixation was done at approximately room temperature. We used a series of switches to transition the cephalon to this circuit without interrupting flow (see Fig. 3).

The washout cart (Fig. 1) consisted of a 2L reservoir of washout solution connected in series with a "T" junction, a peristaltic pump, a Millipore 0.22µm filter, a glass heat exchanger, and a Y-connector, which terminated in a pair of cannulae spaced to match the separation distance between the carotid arteries of a typical rabbit. There was a manometer attached via a side arm directly before the cannulae, which we used to control the pressure during perfusion. There was also a Physitemp MT-23 thermocouple needle probe inserted into the tubing immediately before the carotid cannulae, which we used to monitor perfusate temperature. The cart contained an uninterruptible power supply to enable continuous perfusion of the rabbit cephalon during transport between the operating room and the perfusion room.

The CPA circuit (Fig. 2) consisted of a linear gradient generator, two computer-controlled peristaltic pumps (one for perfusion and one to create the cryoprotectant gradient), and a digital manometer. For additional details on the principles of using a recirculating gradient generator to perfuse organs with cryoprotectants, see Refs. (Fahy 1980; Fahy 1994).

2.3. Chemicals

We used only pharmaceutical grade or higher chemicals in our washout solutions. A full list of chemicals and suppliers is provided in Table 1.

Table 1. Chemicals used for anesthesia, blood washout solutions, fixatives, and to process samples for electron microscopy

Chemical	Supplier	City	Catalog/ID no.
Xylazine	Lloyd	Shenandoah, Iowa	139-236
Isoflurane	Clipper	St Joseph, Missouri	57319-559-06
Ketamine	Putney	Portland, Maine	26637-411-01
Euthasol	Virbac	Fort Worth, Texas	051311-050-01
Sodium Heparin	Hospira	Lake Forest, Illinois	0409-2720-03
Chlorpromazine HCl	WeWard	Eatontown, New Jersey	0641-1397-35
PBS 10× Concentrate	Fisher	Pittsburgh, Pennsylvania	BP665
$NaH_2PO_4 \cdot 2H_2O$	Sigma	St. Louis, Missouri	04269
$Na_2HPO_4 \cdot 2H_2O$	Sigma	–	30435
Glucose	Fisher	–	D16-10
NaCl	Fisher	–	S671-3
Sodium HEPES	Sigma	–	H7006
$NaHCO_3$	Fisher	–	S233-3
$K_2HPO_4 \cdot 3H_2O$	Sigma	–	P5504
$CaCl_2 \cdot 2H_2O$	Sigma	–	C7902
$MgCl_2 \cdot 6H_2O$	Fisher	–	M35-212

Chemical	Supplier	City	Catalog/ID no.
Hydroxyethyl Starch	Serumwerk	Bernburg, Germany	450/0.7
Glutaraldehyde	Fisher	–	G-151-1
Sodium Azide	Sigma	–	S2002
Sodium Nitrite	Sigma	–	S2252
Sodium Dodecyl Sulfate	Sigma	–	L3771
Ethylene Glycol	Spectrum	Gardena, CA	E1051
Agar	Sigma	–	05039
Sodium Cacodylate, 0.2M	EMSa	Hatfield, Pennsylvania	11652
$K_4Fe(CN)_6 \cdot 3H_2O$	Sigma	–	P3289
OsO_4 (4% Solution)	EMS	–	19150
Uranyl Acetate Dihydrate	EMS	–	22400
Absolute Ethanol	EMS	–	15055
Propylene Oxide	EMS	–	20401
SPURS Embedding Kit	EMS	–	14300

A= Electron Microscopy sciences

2.4. Blood washout solutions

Blood washout solutions (Table 2, Table 3) were adjusted to pH 7.40 with HCl and filtered using a Millipore 0.22μm filter. We oxygenated blood washout solutions by bubbling O_2 for at least 1h before use. The washout solutions were oxygenated at room temperature and cooled using the heat exchanger described in Fig. 1.

Table 2.
Composition of the PBS (phosphate-buffered saline) blood washout solution.

Chemical	Concentration
PBS 10× Concentrate	100 mL/L
Ketamine	5.40 mL/L
Sodium Heparin	0.50 mL/L
Euthasol	0.35ml/L

Table 3.
Composition of KR8H: Krebs–Ringer's based blood washout solution.

Chemical	Concentration
Hydroxyethyl Starch	60.00 g/L
NaCl	7.59 g/L
Glucose	1.8 g/L
Sodium HEPES	1.30 g/L
$MgCl_2 \cdot 6H_2O$	1.017 g/L
$CaCl_2 \cdot 2H_2O$	0.169 g/L
$K_2HPO_4 \cdot 3H_2O$	0.51 g/L
$NaHCO_3$	0.42 g/L
Ketamine	5.40 mL/L
Sodium Heparin	0.50 mL/L
Chlorpromazine HCl	0.432 mL/L

For some experiments, we used a modified Krebs–Ringer's solution (KR8H) (Table 3) as a blood washout solution instead of the PBS washout solution.

We also occasionally included 10g/L sodium nitrite in the PBS washout solution as a vasodilator (Palay et al. 1962).

2.5. Fixative solution

The fixative solution was a standard 3% w/v glutaraldehyde solution in a 0.1M phosphate buffer (Table 4; Hayat 1991; Kalimo and Pelliniemi 1977).

Table 4.
Fixative formula.

Chemical	Concentration
$Na_2HPO_4 \cdot 2H_2O$	14.65 g/L
$NaH_2PO_4 \cdot 2H_2O$	2.76 g/L
Glutaraldehyde	3% w/v

The fixative solution was adjusted to pH 7.40 with HCl and filtered with a Millipore 0.22μm filter. The osmolarity of the phosphate buffer was 238 mOsm, and the osmolality of the entire fixative solution was 620 mOsm, as measured by a Precision Systems "Osmette A" freezing point depression osmometer. We found in these studies that "biological grade" glutaraldehyde was just as effective in giving good ultrastructural preservation as the more expensive "electron microscopy grade" glutaraldehyde (Robertson and Schultz 1970), and we therefore used "biological grade" glutaraldehyde for the vast majority of our studies.

2.6. Fixative additives

We also used small quantities of sodium dodecyl sulfate (SDS) to prevent brain shrinkage and sodium azide to prevent mitochondrial swelling (Table 5).

Table 5.
Fixative additives—these chemicals were present in the first 2L of fixative which was perfused in an open circuit at the start of fixation.

Chemical	Concentration
Sodium Dodecyl Sulfate	0.01% w/v
Sodium Azide	0.1% w/v

2.7. Cryoprotectant solution

Our cryoprotectant solution (Table 6) was the same as the fixative solution, except it also contained 65% w/v ethylene glycol. We filtered this solution with a Millipore 0.22μm filter, but did not adjust the pH.

Table 6.
Cryoprotectant solution.

Chemical	Concentration
$Na_2HPO_4 \cdot 2H_2O$	14.65 g/L
$NaH_2PO_4 \cdot 2H_2O$	2.76 g/L
Glutaraldehyde	3% w/v
Ethylene Glycol	65% w/v

2.8. Surgery—bilateral carotid cannulation

Anesthesia was induced in rabbits with an injection of 50mg/kg ketamine and 5mg/kg xylazine. Rabbits were kept at an appropriate surgical plane of anesthesia using 1%–5% isoflurane with 100% O_2 by mask.

Bilateral carotid cannulation was performed under anesthesia as follows: An incision was made in the neck, and the carotids were exposed and dissected free of surrounding tissue. Ligatures were looped loosely around both common carotid arteries. Trickle flow through one carotid cannula was then begun while the contralateral carotid line remained clamped. A nick was made in one common carotid, and the low-flow cannula was immediately inserted and full perfusion with oxygenated washout solution immediately instituted at 80mmHg. Within 30s the ipsilateral jugular vein was severed and the carotid was firmly ligated onto the perfusing cannula. Then the clamp on the contralateral carotid cannula was released, the second carotid was cannulated, and the contralateral jugular vein was cut. Cannulation of both carotids took approximately 2min and did not result in any cerebral ischemia since the Circle of Willis remained perfused through at least three pathways at all times (two vertebral arteries and at least one carotid artery).

Washout continued for 5min–10min at 10°C after the completion of cannulation. The rabbit cephalon was then surgically separated from the body and rolled on the washout cart (Fig. 1) to the CPA circuit (Fig. 2) in the fume hood while perfusion continued.

2.9. Initial fixation

After moving the cephalon to the fume hood as described above, we attached a reservoir containing 2L of fixative solution (Table 4) with SDS and sodium azide additives (Table 5) to the washout circuit via the "T" junction. We perfused this initial fixative at room temperature and 80mmHg in an open circuit until the level of the fixative reservoir reached 200mL, which took 10min–20min depending on the flow rate of the cephalon.

2.10. Transfer to computer control

While perfusing the initial fixative solution, we replaced the manometer with a digital manometer and prepared the perfusion system as detailed in Fig. 3. Once the initial fixative reservoir reached 200mL, we gradually decreased the speed of the cart's peristaltic pump, and the computer-controlled pump (Cole Parmer 07550-30 under National Instruments LabView version 10.0.1 software control) gradually increased to keep the perfusion pressure at 80mmHg. At no point during this transfer did the brain experience any interruption of flow. Once transfer to computer control was complete, we moved the cephalon to the gradient generator and began recirculating fixative without additives (Table 4, Fig. 2). The gradient generator contained 1.5L of fixative solution and 1.5L of CPA solution for a total of 3L.

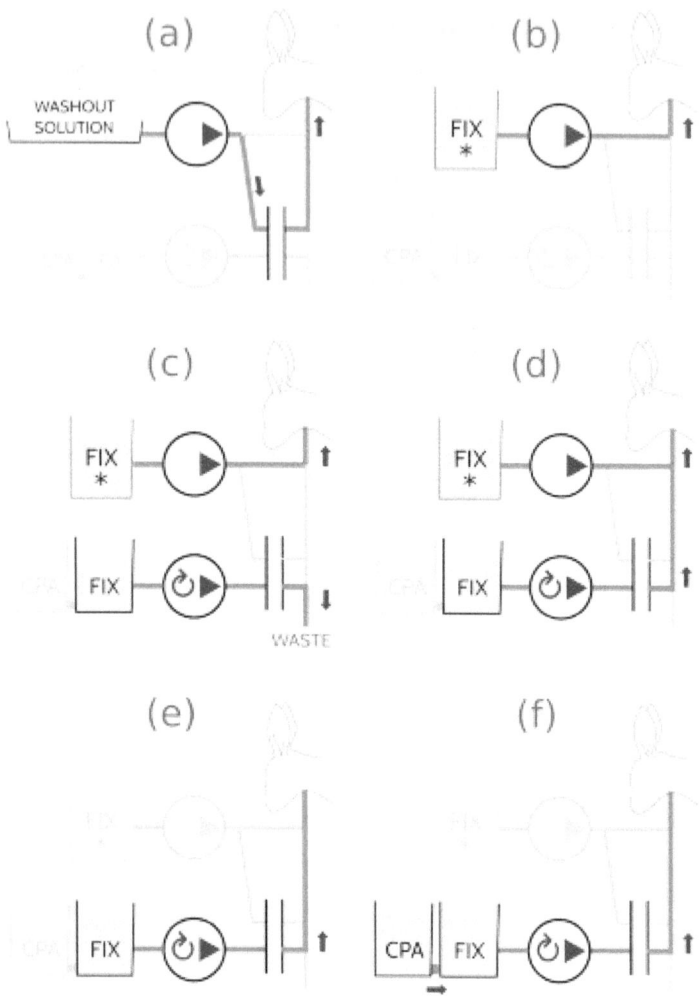

Fig. 3.
Initial fixation was accomplished by replacing the container of washout solution (a) with a container of 2 L of fixative with additives (FIX*) and bypassing the filter (b) to avoid fixative dilution by fluid in the filter. While this initial fixative perfusion was underway, we cleared the filter of blood washout solution by perfusing it with fixative in parallel using the computer-controlled pump (c). Transfer to computer control was accomplished by joining the two pathways (d) and gradually decreasing the speed of the manual pump, causing the computer to increase the rate of the computer-controlled pump to compensate, without disrupting flow to the cephalon (e). After 45 min of total fixation time, we started the CPA ramp (f).

2.11. Cryoprotectant introduction

After 45min total fixative perfusion time, we engaged the gradient generator and started a gradual linear increase in CPA concentration. The CPA ramp took 4h total, during which the computer decreased the perfusion pump speed to maintain perfusion pressure at 80 mmHg as the viscosity of the solution increased. After exhausting the solutions in the gradient generator, we recirculated 500mL of fresh CPA solution for 1h to ensure complete equilibration of cryoprotectant. After this 1h CPA plateau we discontinued perfusion and removed the cephalon from the perfusion machine.

2.12. Changes in cephalon temperature during surgery, fixation, and CPA introduction

Rabbit cephalons remained at body temperature (37°C) during surgery. The perfusion temperature decreased over the course of around 5min to 20°C once we introduced the cooled washout solution. When fixation began, the cephalic temperature was typically around 12°C. With the introduction of room-temperature fixative solution, the perfusion temperature rose to room-temperature (22°C) within 2min–3min and remained at room-temperature throughout the cryoprotectant gradient and cryoprotectant plateau.

2.13. Vitrification

Cryoprotected rabbit cephalons were placed in a 21st Century Medicine, Inc. Controllable Isothermal Vapor Storage (CIVS) device, which maintained temperature at −140°C. Rabbit cephalons were stored in the CIVS for at least 48h and in some cases for several weeks. Temperature was monitored by a Physitemp MT-23 thermocouple needle probe inserted into the rabbits' pharynxes. It typically required 12h for rabbit cephalons to reach −135°C as measured by this thermocouple. We kept rabbit cephalons in the CIVS unit for at least 48h before further processing.

2.14. Application of ASC to porcine brains

For pigs, we used a larger and simpler version of the rabbit perfusion machine, which did not contain a heat exchanger or initial fixative reservoir and which had manual instead of computer-controlled pumps.

As in the case of rabbit experiments, bilateral carotid cannulation was performed while avoiding interruption of perfusion to the brain during blood washout as described above. Blood washout was performed at room temperature instead of at 10°C, again using an oxygen-saturated perfusate. All solutions (blood washout, fixative, and cryoprotectant) were as described previously.

The fixation and cryoprotection schedules for pig brains were the same as for rabbit brains: a 45-min fixation period followed by a 4h linear CPA ramp followed by recirculation of the full-concentration CPA solution for 1h. We maintained the perfusion pressure a bit higher than in the case of rabbits (80–100mmHg) during the CPA ramp owing to the modestly higher vascular resistance of larger organs. We used 8L total fixative in the gradient generator—4L fixative and 4L CPA solution. We used 1L CPA solution during the CPA plateau. Because the pig cephalon perfusion machine did not have a separate initial fixative reservoir, we included the sodium azide and SDS additives at full concentration in the entire 4L fixative solution in the gradient generator.

Pig cephalons were stored at least one week in the CIVS unit. We found that pig cephalons reached −135°C after 24h in the CIVS, as measured by a needle thermocouple probe deeply inserted into the pig's pharynx via the snout.

2.15. Warming and removal of CPA

After vitrification, we removed the cephalon (pig or rabbit) from the CIVS unit and allowed it to warm to room temperature in a fume hood for 2h. We removed CPA from rabbit brains using either perfusion removal from rabbit cephalons or diffusion removal from rabbit brain slices. We removed CPA from pig brains using diffusion from brain slices only.

For perfusion removal of CPA from rabbit brains we used essentially the reverse of the process that was used to load CPA into the rabbit brain: We reattached the rabbit cephalon to the computer-controlled perfusion machine (Fig. 2) and removed CPA via a slow linear decrease over 4h using the gradient generator. The computer increased the flow rate to keep the pressure constant at 80mmHg as the viscosity of the perfusate decreased. After exhausting the solutions in the gradient generator (1.5L of fixative solution and 1.5L of CPA solution for a total of 3L), we recirculated 500mL fresh fixative solution for 1h to ensure complete removal of cryoprotectant.

After removing CPA via perfusion, we disconnected the cephalon from the perfusion machine, extracted the brain from the skull, embedded it in agar, and cut 150μm coronal slices on a Pelco 100 vibratome. We placed the slices in small vials filled with 0.1M cacodylate buffer (pH 7.4, 200mOsm).

For diffusion removal of cryoprotectant, we extracted the CPA loaded brain from the skull, embedded it in agar containing 65% w/v ethylene glycol, and cut 150μm coronal slices on the vibratome, which we placed in small vials filled with CPA solution. We removed the CPA from the brain slices via exponential dilution with 0.1M cacodylate, replacing half of the CPA solution surrounding the slices with 0.1M cacodylate once every 10min. We considered the CPA to be removed after completing five rounds of exponential dilution (to a final concentration of approximately 3% w/v ethylene glycol).

For porcine brains, we cut the brain into 3.2 cm × 1.2 cm × 2 cm blocks using a Thomas Scientific tissue slicer blade (catalog #6727C18), embedded those blocks in agar containing 65% w/v ethylene glycol, and cut vibratome slices and removed CPA as described for rabbit brain slices.

2.16. Staining and embedding brain tissue for electron microscopy

After removing the CPA, we used the Focused Ion Beam Milling and Scanning Electron Microscopy (FIB-SEM) protocol for staining and dehydration as described by Graham Knott et al. We washed the brain slices three times with 0.1M cacodylate for 5min each, then stained with 1.5% w/v potassium ferrocyanide and 1% w/v osmium tetroxide in 0.1M cacodylate buffer for 30min. Next, we stained slices with 1.0% w/v osmium tetroxide in 0.1M cacodylate buffer for 30min. We washed the brain slices with double distilled water for 3min and then stained them with 1% w/v uranyl acetate solution for 30min. Then we washed the brain slices with double distilled water for 5min and dehydrated the slices in a graded ethanol series: 50% ethanol for 2min, 70% ethanol for 2 min, and finally four successive rinses with 100% ethanol for 2min each.

We embedded brain slices in SPURS resin using 3 rounds of 100% propylene oxide (PO) for 3min each, followed by a 1:1 mixture of SPURS resin and PO for 30min, then 1:2 PO:SPURS for 30min, then finally 100% SPURS for 12h overnight. We transferred the samples to fresh SPURS resin and polymerized the resin in a Quincy Lab Inc. Model GC laboratory oven at 65°C for 48h.

After hardening we cut 0.33μm thick sections for histology using a Reichert-Jung Ultracut-E ultramicrotome, stained those sections with toluidine blue, and viewed them under a Nikon eclipse 50i light microscope using Nikon Plan Achromat 4×, 10×, 20×, and 40× objective lenses.

For electron microscopy, we cut 80nm thin sections on the ultramicrotome using a Diatome diamond knife, then coated the sections with a 10nm layer of carbon in a Bal-Tec SCD 500 carbon coater. We viewed sections on a Zeiss Supra 40VP FESEM using a STEM detector at 28kV.

2.17. Controls

We controlled for the effects of cryoprotectant introduction and vitrification by preparing rabbit brains using the same surgical procedures, perfusion machines, and chemicals as with ASC preserved brains, except that after 45min total fixative perfusion we did not begin the CPA gradient. Instead we transferred the rabbit cephalon to a refrigerator at 4°C for at least 4h, then removed the brain from the skull and performed staining, dehydration, embedding, and analysis as described above.

3. Results

3.1. Preservation of rabbit brains

Of the 37 rabbits used during the course of these experiments, 29 were used to refine the parameters of the ASC protocol and establish controls and 8 were processed using the ASC protocol described previously. Of these 8 rabbits processed for ASC, 3 were processed with KR8H washout solution, 3 were processed with PBS washout solution, and 2 were processed with PBS washout solution with 10g/L sodium nitrite included.

Gross observation of all 8 ASC processed rabbit brains upon dissection revealed no cracks resulting from the vitrification or rewarming processes. Brain weights were commensurate with control brains, and we found no retraction of the brains from their skulls. Control rabbit brains displayed excellent ultrastructural preservation, as expected. Fig. 4 shows control CA1 hippocampal neurons at relatively low magnification (2,420×).

All 8 rabbit brains preserved using ASC consistently displayed ultrastructural preservation indistinguishable from that of controls, as indicated in Fig. 5 showing ASC-preserved pyramidal neurons also from the CA1 band of the rabbit hippocampus at the same magnification. In both electron micrographs, uniformly intact cell membranes are clearly visible and cells appear clear. The absence of "dark" neurons (Cammermeyer 1961), indicates proper fixation and the lack of mechanical disruption to brain tissue. Nuclear envelopes are clearly defined and display no discontinuities. Intracellular organelles are also well preserved: rough endoplasmic reticulum is clear and compact, and the mitochondria appear normal, with cristae clearly visible even at the relatively low magnification of the image. There are no visible examples of exploded and vacuolated (or "popcorned" Minassian 1979) mitochondria. The few darkly-stained myelinated transverse processes seen are well-preserved, with tight myelin sheaths.

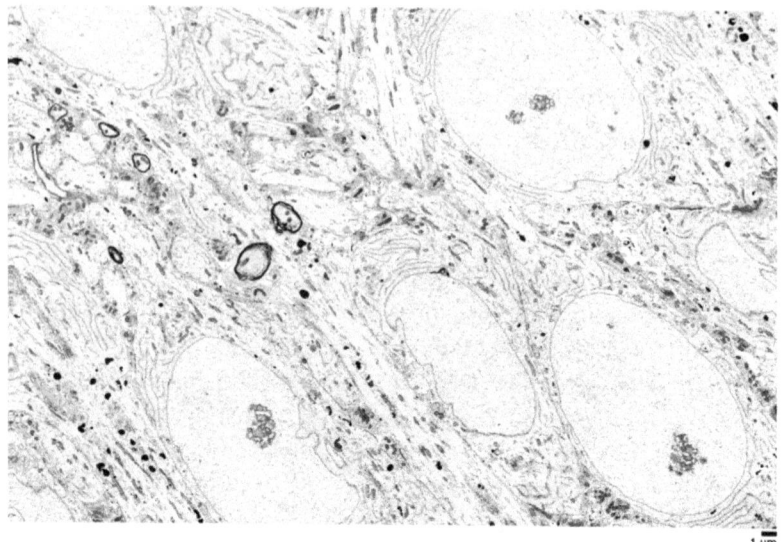

Fig. 4.
Cells of the CA1 band of rabbit hippocampus. Control: Brains were washed out with PBS washout solution, fixed for 45min, then stained and embedded. Experiment date: 2015-05-27. 2,420×.

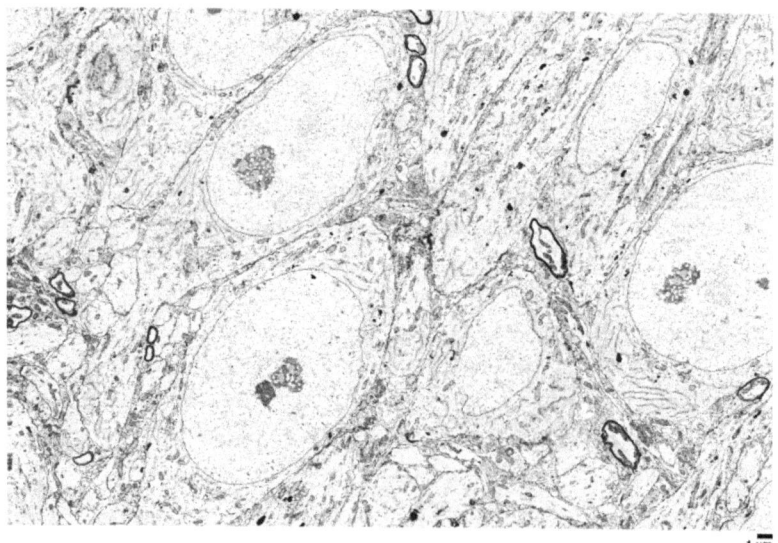

Fig. 5.
Cells of the CA1 band of rabbit hippocampus. KR8H washout solution. Vitrified; CPA removed by perfusion. Experiment date: 2015-04-21. 2,420×.

Fig. 6 is a high magnification (15,500×) shot of neuropil near the CA1 band of a control rabbit hippocampus, and Fig. 7 shows a comparable area in an ASC-preserved brain. Both images show a single myelinated axon and several other unmyelinated synaptic processes. Each process is unambiguously defined by crisp membranes. There are several synapses present, with clear pre-synaptic vesicles and well defined, darkly stained post-synaptic densities. At this level of magnification, the cristae of the mitochondria are readily apparent. The large process in the upper right of the image displays clear neurofilaments and a clear cytoskeleton; a transverse view of the cytoskeleton can be seen in the neural process in the center of the image.

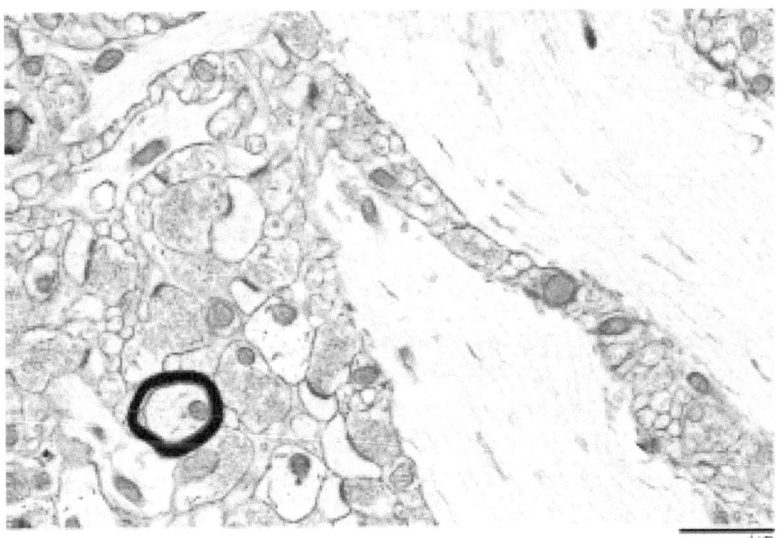

Fig. 6.
Rabbit brain. Neuropil near the CA1 band in the hippocampus. Control: Brains were washed out with PBS washout solution, fixed for 45min, then stained and embedded. Experiment date: 2015-05-27. 15,500×.

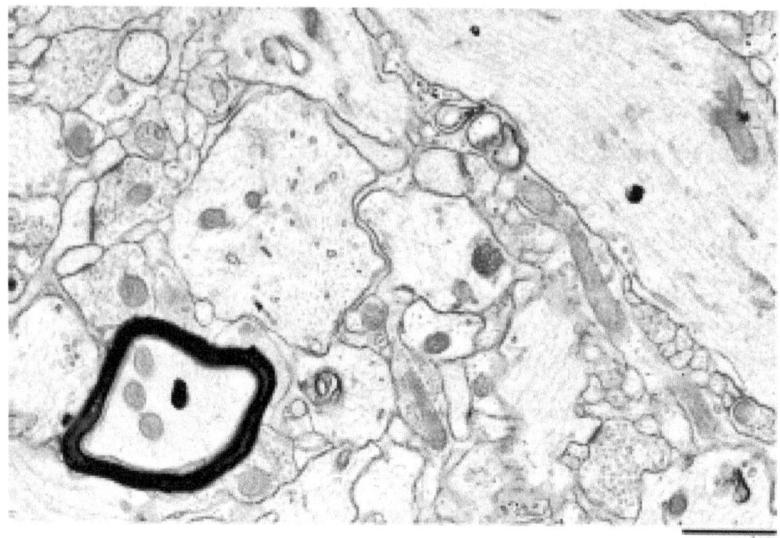

Fig. 7.
Rabbit brain. Neuropil near the CA1 band in the hippocampus. Synapses, vesicles, and microfilaments are clear. The myelinated axon shows excellent preservation. KR8H washout solution. Vitrified; CPA removed by diffusion. Experiment date: 2015-04-15. 15,500×.

Perfusion removal of CPA resulted in ultrastructural preservation equivalent to diffusion removal of CPA: The brain shown in Fig. 7 CPA removed via perfusion while the brain shown in Fig. 5 had CPA removed via diffusion. There is no appreciable difference between the quality of preservation attained in these two cases, or in other similar comparisons not shown.

To demonstrate the suitability of ASC brains for connectome tracing studies, Ken Hayworth at the Janelia Research Campus in Virginia kindly agreed to carry out FIB-SEM imaging of a 10μm × 10μm × 5μm region of rabbit hippocampus that had been preserved using ASC. Fig. 8 shows several slices from that volume. Overall structural preservation is excellent: processes are clearly defined and organelles are intact, as expected. When observing slices of this volume in sequence, it is easy to track the progression of any process through the stack, demonstrating that connectivity in this region was not impaired by our preservation method (see full video available in online supplemental materials).

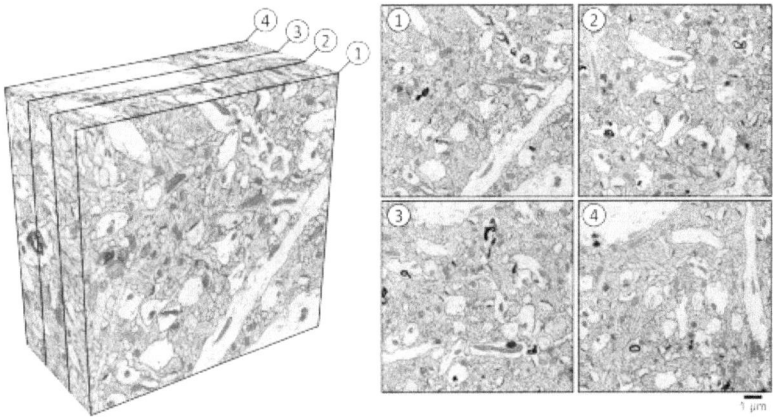

Fig. 8.
Frames from a FIB-SEM stack of rabbit neuropil near the CA1 band of the hippocampus. Stack dimensions: 10μm × 10μm × 5μm. All processes appear well preserved and traceable throughout the stack. Full video available in supplemental materials. KR8H washout solution. Vitrified; CPA removed by diffusion. Experiment date: 2015-04-15. FIB-SEM done by Ken Hayworth at Janelia Research Campus. Used with permission.
Supplementary data related to this article online can be found at http://dx.doi.org/10.1016/j.cryobiol.2015.09.003.

Figs. 9 and 10 further demonstrate the level of preservation that can be obtained using ASC. Fig. 9 shows the edge of a CA1 pyramidal cell from the rabbit hippocampus. Nuclear pores and a Golgi apparatus are visible. Two nodes of Ranvier are also visible near the center of the image. Fig. 10 shows detailed, classic "clasping hand" synapses, as well as one synapse with three post-synaptic densities. Pre-synaptic neurotransmitter vesicles are clearly defined.

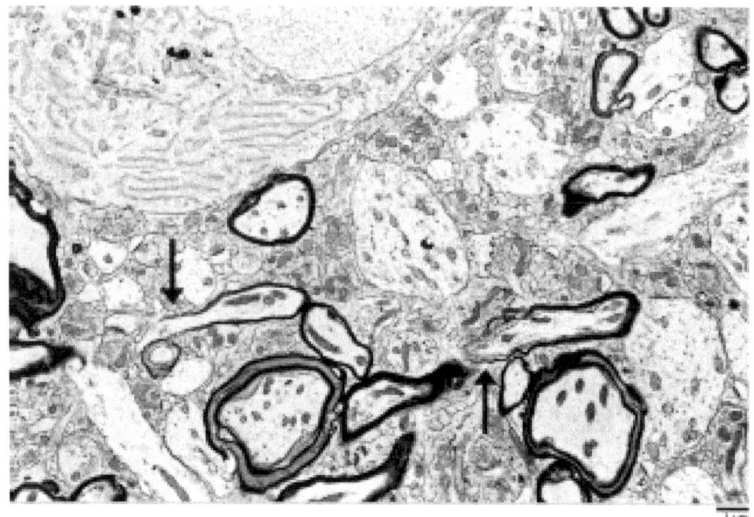

Fig. 9.
Excellent preservation of myelinated and unmyelinated axons near a CA1 cell in rabbit hippocampus. Cell organelles and endoplasmic reticulum appear normal. Note the nodes of Ranvier (arrows) in the middle of the image. PBS washout solution. Vitrified; CPA removed by diffusion. Experiment date: 2015-04-22. 5,450×.

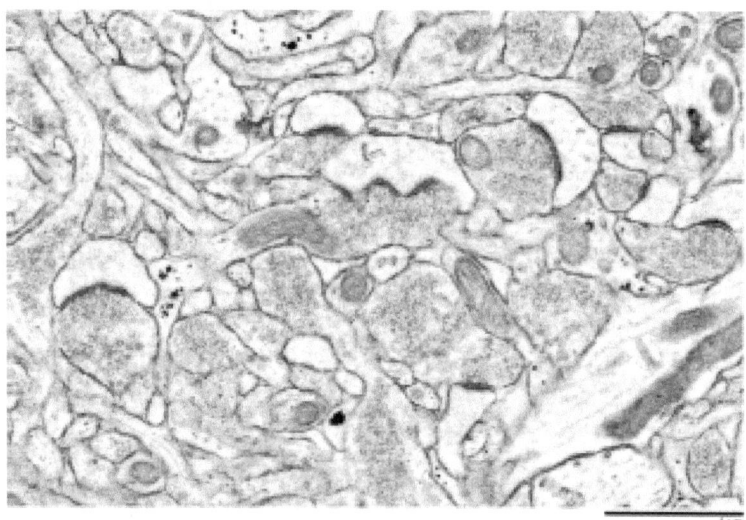

Fig. 10.
Neuropil near CA1 band of rabbit hippocampus. High magnification image of multiple synapses showing clear pre-synaptic vesicles and post-synaptic densities. PBS washout solution with 10g/L sodium nitrite. Vitrified; CPA removed by diffusion. Experiment date: 2015-05-06. 23,130×.

We also investigated preservation in the cerebellum and thalamus and found acceptable preservation in both cases. Fig. 11 shows ASC preserved rabbit cerebellum. All cells appear well preserved, and the two capillaries shown display good integrity, are not collapsed, and are free of debris. There is some slight loosening of the myelin sheath surrounding one of the heavily myelinated processes, but the process itself is still well defined, which would enable uninterrupted connectomic analysis.

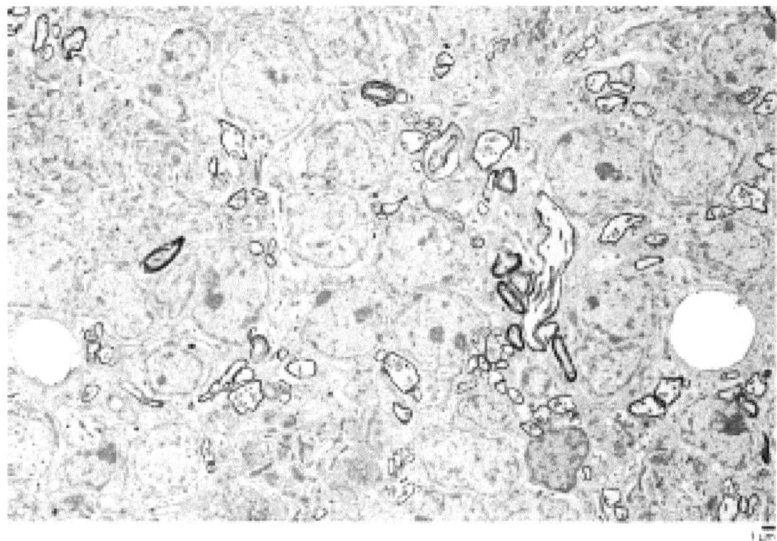

Fig. 11.
Rabbit cerebellum. Overall structure appears good; myelinated and unmyelinated processes are clear. PBS washout solution with 10g/L sodium nitrite. Vitrified; CPA removed by diffusion. Experiment date: 2015-05-06. 2,180×.

Fig. 12 shows preservation in rabbit thalamus. Here there is some apparent damage to the myelin sheaths of the heavily myelinated tracts of axons. However, even in the presence of this damage, all processes appear to us to be traceable and intact, and thus suitable for connectomic analysis.

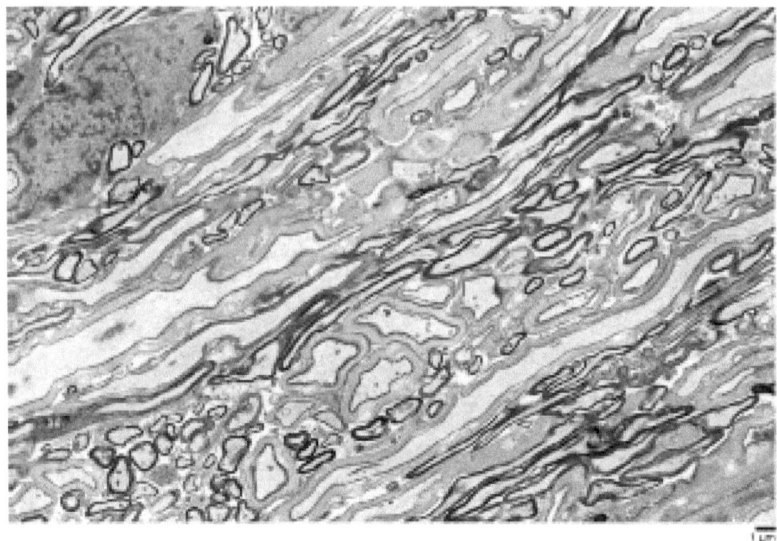

Fig. 12.
Rabbit thalamus. There is some distortion of the heavily myelinated axons, though all processes appear traceable. PBS washout solution with 10 g/L sodium nitrite. Vitrified; CPA removed by diffusion. Experiment date: 2015-05-06. 3,310×.

In general, the ultrastructure in each of these figures displays the lack of extracellular space which is typical of perfuse-fixed brain tissue prepared for electron microscopy (Cragg 1980; Hayat 1991).

3.2. Preservation of pig brains

Of the three pigs used during the course of these experiments, surgical complications led to two pigs suffering ischemic time in excess of 15min (micrographs not shown). The remaining pig was successfully processed using ASC as described previously.

Fig. 13 shows a histology image of pig hippocampal CA1 pyramidal neurons and surrounding cortex. This entire region of brain tissue displays excellent preservation: All capillaries are open and clear of debris, there are no "dark" cells, and there is no obvious mechanical or osmotic disruption or distortion of any cells.

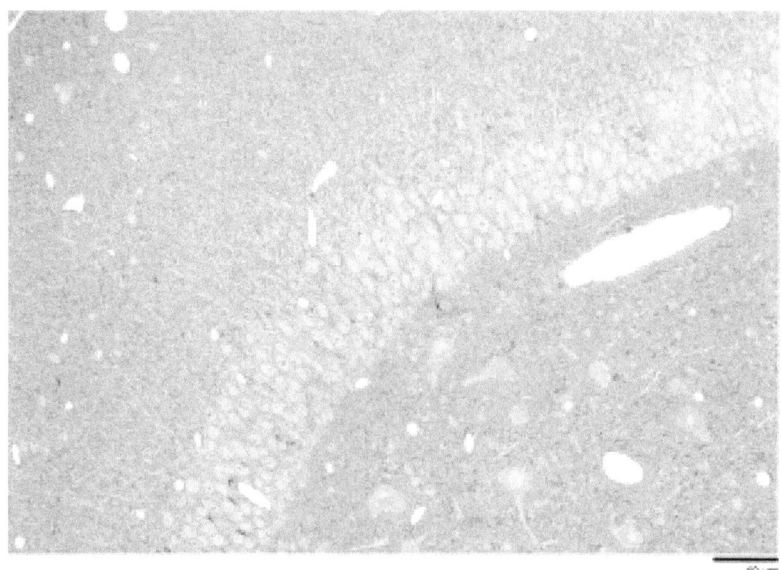

Fig. 13.
Pig hippocampus CA1 band and cortex. Note the excellent histology of all cells present. 0.33 µm section stained with toluidine blue. PBS washout solution. Vitrified; CPA removed by diffusion. Experiment date: 2015-04-28. 20×.

Fig. 14 is an image of pig neuropil near the CA1 hippocampal band and is comparable to the rabbit micrographs shown in Figs. 8 and 9. As in those images, there are multiple well-preserved synapses, mitochondria appear normal, neurofilaments appear undamaged, and all processes are clearly defined by well-preserved membranes.

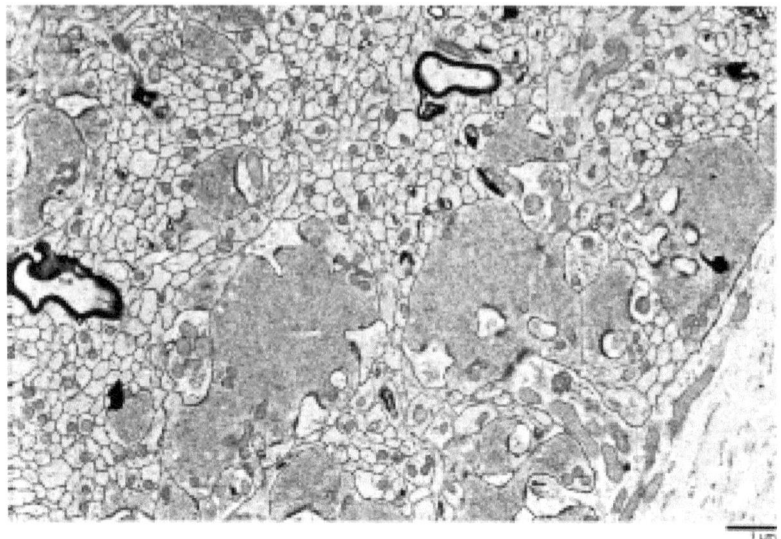

Fig. 14.
Neuropil of pig brain near the CA1 area of the hippocampus. Note the large "pools" of vesicles. PBS washout solution. Vitrified; CPA removed by diffusion. Experiment date: 2015-04-28. 8,050×.

In general, there was no difference in the quality of preservation obtained using pig or rabbit brains. Additional pig brain images as well as a FIB-SEM video are available in the online supplemental materials.

3.3. General methodological observations

We did not see any difference in preservation quality between KR8H blood washout solution (Table 3) (shown in Fig. 5, Fig. 7 and Fig. 8) and the simpler PBS-based blood washout solution (Table 2) (shown in Figs. 4–6 and Fig. 8–10 and 13 and 14).

For some experiments such as those shown in Figs. 11 and 12, we employed 10gL sodium nitrite in the washout solution as a vasodilator (Palay et al. 1962). We found no difference in the quality of preservation obtained with or without sodium nitrite.

On the other hand, we found that 0.1% w/v sodium azide added to our initial fixative solution completely eliminated "popcorn mitochondria" in our electron micrographs, which we often observed without the azide. Sodium azide has been used previously to prevent mitochondrial swelling during immersion fixation (Minassian and Huang 1997).

4. Discussion

4.1. Brain cryoprotection, dehydration prevention, and vitrification

We chose ethylene glycol as the sole cryoprotectant in our CPA solution (Table 6) based in part on permeability studies we conducted on fixed brain slabs which indicated that ethylene glycol was the most permeable of all CPAs studied. It would be reasonable to assume that fixation would permeabilize both the blood-brain barrier and the brain itself to cryoprotectants, but we rapidly learned that this was far from the case: we observed gross brain shrinkage and shrunken and dehydrated myelin processes, even with very extended (80h) cryoprotectant introductions and very low (50mOsm) phosphate buffer osmolalities (images not shown) even using the most permeable CPA available. We therefore employed SDS in view of the fact that SDS can be used to reversibly permeabilize the blood–brain barrier in rats (Saija et al. 1997), and has previously been used to accelerate cryoprotectant uptake in unfixed tissue (Yuri Pichugin, personal communication). SDS was found to be critical for our purposes by allowing cryoprotectant to penetrate the brain without causing shrinkage. When SDS was included, we found no observable brain shrinkage when measuring brain weight or examining ultrastructure.

We are confident that both small and large ASC brains processed as described vitrify when cooled. 65% w/v ethylene glycol is 10% w/v more concentrated than a concentration that will vitrify at a cooling rate of about 10 °C/min (Fahy 1988), and should have an extremely low critical cooling rate. Most of our brains approached −140°C prior to rewarming, and all descended to at least −135°C, whereas the glass transition temperature of 65% w/v ethylene glycol even when a carrier is absent is close to −131°C (Fahy and Wowk 2015). We also observed no signs of ice crystal artifacts in any of our ASC-processed brains.

To further demonstrate vitrification of our cryoprotectant solution, we prepared a sample of our cryoprotectant solution and stored it in the CIVS for 12h, monitoring temperature with the same needle thermocouple we used for the rabbit cephalons. After removing the sample of cryoprotectant after it had reached −135°C, we found that it had formed a solid glass, without crystals.

Vitrified storage at −135°C should enable essentially indefinite storage of brain tissue with no degradation due to suppressed molecular motion in the vitrified state.

4.2. Brain banking for the 21st century

Aldehyde-stabilized cryopreservation promises to be a superior brain banking technique compared to other methods. ASC can enable preservation of precious samples such as brains from expensive or extensively modified research animals. These samples can then be analyzed by multiple labs and by multiple techniques that are compatible with previous aldehyde fixation, without constraints on sample storage time. Current fixation methods do not allow indefinite preservation (Hooshmand et al. 2014; Xie et al. 2011), a problem that ASC is believed to solve, given that the standard principles of cryopreservation (Fahy and Rall 2015, Fahy and Wowk 2015) apply to ASC-preserved brains. Thus, we believe the current studies show that the principles and methods of cryobiology in general and of organ vitrification in particular have now been demonstrated to have applications in a new area of biological research.

We studied the hippocampus in part because it is a brain region that is particularly sensitive to ischemic injury (Ruan et al. 2009) and thus in principle a particularly delicate and challenging structure to preserve. It is also essential for the formation of long-term memories (Tsien et al. 1996), and CA1 cells in particular pass information from the hippocampus to other sites in the brain (Anderson et al. 2007). Therefore, the results obtained for CA1 cells in the present work are particularly encouraging for demonstrating the ability of ASC to preserve both delicate and complex brain structures.

We have shown that cryoprotectant can be removed either by simple diffusion (Fig. 7) or by reperfusion for gradual removal (Fig. 5) Perfusion removal of CPA enables a smooth transition to any technique that can start with an aldehyde-fixed brain (even ones that require subsequent perfusion with different chemicals), and is appropriate when CPA must be removed from the whole brain prior to analysis.

We envision that ASC brains might be subsequently embedded using a whole brain staining and embedding protocol such as Mikula's BROPA protocol (Mikula and Denk 2015), or used with completely different whole brain perfusion based protocols such as the CLARITY method (Chung et al. 2013), expansion microscopy (Chen et al. 2015), or other methods yet to be devised. Preserved brain slices can also be taken and shipped to multiple labs for analysis. Collaborative research of this kind should extend access to complex analytic techniques, enabling a single lab with access to unique material to stockpile brains and then distribute them to multiple analytical sites, and even to sites that did not exist at the time of stockpiling.

We have shown that both rabbit brains (10g) and pig brains (80g) can be preserved equally well. We do not anticipate that there will be significant barriers to preserving even larger brains such as bovine, canine, or primate brains using ASC. We do not, however, necessarily see easy application of our method in current human brain banks (Kretzschmar 2009), because these banks normally receive donated brains many hours after death, which we assume will make them difficult to adequately perfuse. Furthermore, we do not see any simple way for ASC to be compatible with research protocols which require both fixed and unfixed tissue from the same brain; ASC delivers chemicals via perfusion and so must fix and cryopreserve the entire brain.

Although we have emphasized the neurobiological applications of ASC here, our method is general and should be applicable to any organ system, or even to entire animals. The National Institute on Aging, for example, maintains stocks of aged animals—a transient and precious resource—and samples might be preserved for later analysis by gerontologists. In another instance, human organs provided to scientists by organizations such as the

National Disease Research Interchange might also be processed using ASC to ensure that organs suitable for morphological studies are always available and not needlessly wasted. Finally, although we have used glutaraldehyde for preserving ultrastructure, other fixatives can also be used with ASC to enable, for example, better detection of epitopes that might be altered by glutaraldehyde.

In the present work, we've shown the development of a new and general brain banking technique for connectomics and other types of neuroanatomical research that should enable valuable material to be preserved without time constraints and shared between labs for comprehensive examination. ASC promises to be a powerful new technique in the quest of connectomics researchers to unravel the mysteries of the mind. We also hope that ASC will inspire investigators in other fields to consider the possibility of cryobiological solutions to their problems.

Conflicts of interest

None.

Acknowledgments

This research was supported by 21st Century Medicine and by the Brain Preservation Foundation.

We thank Brian Wowk for providing perfusion engineering and thermodynamics advice, programming the LabView computer pressure control interfaces, and help with editing of the manuscript. We further thank Dylan Holmes for creating the perfusion machine diagrams and extensive editing help, Bruce Thomson for his excellent electron microscope work and support, and Ken Hayworth for generously providing FIB-SEM analysis of our brain samples. Annemarie Southern and Anita Williams prepared many of our solutions, and Victor Vargas preformed the analytical chemistry work for our brain slab permeability studies and spectroscopic studies of the purity of our glutaraldehyde solutions. Victor Vargas and Limdo Chow also helped construct the pig perfusion machine. Last but not least, we are indebted to Xian Ge, Roberto Pagotan, Adnan Sharif, and Angie Olivia for providing necessary animal surgical support and to John Phan for advice on cryoprotectant perfusion techniques.

References

Andersen P, Morris R, Amaral D et al. (2007) The Hippocampus Book. Oxford University Press

Cammermeyer J (1961) The importance of avoiding "dark" neurons in experimental neuropathology. Acta Neuropathol 1: 245–70

Chen F, Tillberg PW, Boyden ES (2015) Expansion microscopy Science 347: 543–8

Chung K, Wallace J, Kim SY et al. (2013) Structural and molecular interrogation of intact biological systems. Nature 497: 332–7

Clark P, Fahy G, Karrow JAM (1984) Factors influencing renal cryopreservation. ii. toxic effects of three cryoprotectants in combination with three vehicle solutions in nonfrozen rabbit cortical slices. Cryobiology 21: 274–84

Cragg B (1980) Preservation of extracellular space during fixation of the brain for electron microscopy. Tissue Cell 12: 63–72

Eisenstein M (2009) Neural circuits: Putting neurons on the map. Nature 461: 1149–52

Fahy G, MacFarland D, Angell C, Meryman H (1984a) Vitrification as an approach to cryopreservation. Cryobiology 21: 407–26

Fahy G, Rall W (2007) Vitrification: an overview. In: Liebermann J, Tucker MJ (Eds.), Vitrification in Assisted Reproduction: a User's Manual and Troubleshooting Guide, Informa, UK, 1–20

Fahy G, Takahashi T, Crane A (1984b) Histological cryoprotection of rat and rabbit brains. Cryo Letters 5: 33–46

Fahy GM (1980) Activation of alpha adrenergic vasoconstrictor response in kidneys stored at −30°C for up to 8 days. Cryo Letters 1: 312–7

Fahy GM (1988) Vitrification. In: McGrath J, Diller K (Eds.), Low Temperature Biotechnology: Emerging Applications and Engineering Contributions, The American Society of Mechanical Engineers, New York 113–46

Fahy GM (1994) Organ perfusion equipment for the introduction and removal of cryoprotectants. Biomed Instrum Technol 28: 87–100

Fahy GM (2010) Cryoprotectant toxicity neutralization. Cryobiology 60: S45–53

Fahy GM, Rall WF (2015) Overview of biological vitrification. In: Tucker MJ, Liebermann J (Eds.), Vitrification in Assisted Reproduction: from Basic Science to Clinical Application, Taylor and Francis Books, Ltd., New York 1–22

Fahyy GM, Wowk B (2015) Principles of cryopreservation by vitrification. Methods Mol Biol 1257: 21–82

Fahy GM, Wowk B, Wu J et al. (2004) Cryopreservation of organs by vitrification: perspectives and recent advances. Cryobiology 48: 157–78

Griffiths G (1993) Fine Structure Immunocytochemistry. Springer-Verlag, Berlin, Heidelberg, New York

Hayat M (1991) Fixation for Electron Microscopy. Academic Press, London, New York

Hooshmand M, Anderson A, Cummings B (2014) Improved pre-embedded immuno-electron microscopy procedures to preserve myelin integrity in mammalian central nervous tissue. In: Mendez-Vilas A (Ed.), Microscopy: Advances in Scientific Research and Education, Formatex Research Center 59–65

Kalimo H Pelliniemi LJ (1977) Pitfalls in the preparation of buffers for electron microscopy. Histochem J 9: 241–6

Knott G, Rosset S, Cantoni M (2011) Focused ion beam milling and scanning electron microscopy of brain tissue J. Vis. Exp. e2588 URL https://www.jove.com/video/2588/focussed-ion-beam-milling-scanning-electron-microscopy-brain

Kretzschmar H (2009) Brain banking: opportunities, challenges and meaning for the future. Nat Rev Neurosci 10: 70–8

Lichtman JW, Sanes JR, Livet J (2008) A technicolour approach to the connectome. Nat Rev Neurosci 9: 417–22

Mikula S, Binding J, Denk W (2012) Staining and embedding the whole mouse brain for electron microscopy. Nat Methods 9: 1198–1201

Mikula S, Denk W (2015) High-resolution whole-brain staining for electron microscopic circuit reconstruction. Nat Methods 12: 541–6

Minassian H, Huang SN (1979) Effect of sodium azide on the ultrastructural preservation of tissues. J Microsc 117: 243–53

Palay SL, Gordon SS, McGee-Russell M, Grillo MA (1962) Fixation of neural tissues for electron microscopy by perfusion with solutions of osmium tetroxide. J Cell Biol 12: 385–410

Robertson EA, Schultz RL (1970) The impurities in commercial glutaraldehyde and their effect on the fixation of brain. J Ultrastruct Res 30: 275–87

Rosene DL, Rhodes KJ (1990) Cryoprotection and freezing methods to control ice crystal artifact in frozen sections of fixed and unfixed brain tissue. Methods Neurosci 3: 360–585

Ruan YW, Lei Z, Fan Y et al. (2009) Diversity and fluctuation of spine morphology in ca1 pyramidal neurons after transient global ischemia. J Neurosci Res 87: 61–8

Saija A, Princi P, Lanza DTM, Pasquale AD (1997) Changes in the permeability of the blood-brain barrier following sodium dodecyl sulphate. administration in the rat. Exp Brain Res 115: 546–51

Seung S (2012) Connectome: How the Brain's Wiring Makes Us Who We Are. Houghton Mifflin Harcourt Publishing, Boston

Sporns O (2014) Contributions and challenges for network models in cognitive neuroscience. Nat Neurosci 17: 652–60

Tsien JZ, Huerta PT, Tonegawa S (1996) The essential role of hippocampal ca1 nmda receptor-dependent synaptic plasticity in spatial memory. Cell 87: 1327–38

Wowk B (2010) Thermodynamic aspects of vitrification. Cryobiology 60: 11–22

Xie R, Chung JY, Ylaya K et al. (2011) Factors influencing the degradation of archival formalin-fixed paraffin-embedded tissue sections. J Histochem Cytochem 59: 356–65

Combined Approach to the Development of Protocol for Vitrification of Bulky Biological Objects

Pulver A., Artyuhov I.V, Artyukhov V.I., Tselikovsky A.V,
Shamaev N.V., Pulver N.A, Peregudov A

Abstract

Any means of cryopreservation generally relies upon vitrification, in which cells survive in glass between ice crystals. Proposals of a simultaneous vitrification of entire samples without the formation of ice even predate the discovery of the first cryoprotectants, in 1940s. However, the first practical implementation of vitrification (murine preimplantation embryos) was carried out only in 1985, by Gregory Fahy and William Rall. The present report provides a brief overview of different glass types resulting from liquids in a living system converting into the glassy state at low temperatures, as well as the potential benefits of vitrification for long-term large-sized sized biological objects preservation. We specify the main outstanding obstacles for the implementation of bulky biological objects vitrification, as well as ways proposed by various researchers to solve some of them.

One of those obstacles is the toxicity of extremely high cryoprotectant concentration (up to 60% v/w) required for vitrification. That's why we turned our attention to xenon (Xe). The main prospective benefit of using Xe as a cryoprotectant is its absolute nontoxicity on tissue and cellular level, while even the most low toxic cryoprotectants (such as glycerol, ethylene glycol etc.) are potentially dangerous due to osmotic effects.

In order to assess the cryoprotective potential of Xe we have conducted two series of experiments—with yeast (S. cerevisiae) and mammalian (CHO-K1 and NIH-3T3) cell cultures in custom-built bronze miniature hyperbaric chambers.

Yeast survival in presence of Xe at all tested pressures appeared to be much better than in control experiments and approximately equal or slightly better than with 5% DMSO or glycerol; in experiments with joint application of Xe and DMSO or glycerol results have been even better. But in case of mammalian cells convincing results have been achieved only in experiments with the fastest cooling achievable.

Molecular dynamics simulations of Xe clathrate hydrate and ice crystal growth indicate vitrification of water inside cells under combined application of Xe and DMSO/glycerol as a possible explanation of the results observed (details are described in our report "Possible Mechanisms of Cryoprotective Effect of Xenon").

Based on this hypothesis we offer approaches to both cooling (combined use of traditional cryoprotectants and clathrate forming gases for vitrification, patent applied), and to the subsequent heating (uniform heating of complex biological objects by means of electric and magnetic fields' phased emitters, guided by MRI thermometry, patent pending—discussed in yet another corresponding report at this conference), providing the theoretical background for the successful cryopreservation of bulky biological objects. And eventually—up to individual organs and whole organisms.

In addition, we offer some suggestions for improving the efficiency of gas perfusion for chilling procedure, and application of frequency and width-pulse electromagnetic field modulation within the initial stages of cooling.

Introduction

Before describing the essence of our propositions, it wouldn't be inappropriate to recall some matters of common cryobiological knowledge.

Any means of cryopreservation generally relies upon vitrification, in which cells survive in glass between ice crystals. Even under slow programmable freezing cells are surrounded by unfrozen liquid between growing ice crystal (Mazur 1984).

Vitrification is a process whereby fluid becomes a solid during cooling (more precisely, exhibits most of the properties of solids) without any substantial change in molecular arrangement or thermodynamic state variables. Hence, no crystallization with the consequent heat release and volume increase. Proposals of a simultaneous vitrification of entire samples without the formation of ice Luyet 1937 even predate the discovery of the first cryoprotectants, (Polge et al. 1949 in 1940). However, the first practical implementation of vitrification (murine preimplantation embryos) was carried out only in 1985, by Gregory Fahy and William Rall (Rall and Fahy 1985).

The main condition for the onset of vitrification during cooling is an increase of liquid viscosity up to about 10^{13} Poise. On the molecular level it is expressed in loss of rotational and translational degrees of freedom, leaving only bond vibration within a fixed molecular structure, which leads to a decrease in the specific heat and thermal expansion coefficient (Wowk 2010).

Cryoprotectants lower the critical cooling rate (which is not less than 10^7 °C/min for pure water, Armitage 1991) required for vitrification inversely proportional to their concentration within the cooling solution, reducing it up to quite available speeds.

Cryopreservation of individual cell suspensions requires nothing more than a slow programmable freezing (SPF), and the vitrification protocols for whimsical preimplantation embryos (Isachenko et al. 2003) are even more easy to apply. However, when dealing with macroscopic objects, researchers are faced with a number of additional obstacles, namely (Fahy et al. 2009; Mazur, 2010; Hopkins, Badeau et al., 2012):

1. The toxicity of extremely high cryoprotectant concentration (up to 60% v/w) required for vitrification.
2. Uneven cryoprotectants distribution in organ parenchyma.
3. Irregularity of cooling (temperature gradients).
4. Insufficient cooling rate.
5. Cracking during storage below glass transition temperature (T_g).
6. Occurrence of nucleation sites at about-Tg temperatures.

7. Insufficiently slow and uneven warming with temperature gradients and local overheating.
8. Devitrification during warming due to insufficient heating rate with ice recrystallization.
9. Reperfusion syndrome at the end of rewarming.

Ice formation during warming happens faster than during cooling because ice nucleation occurs at lower temperatures than ice growth. This nucleation leads to extensive ice growth at warmer temperatures. That's why the "critical warming rates" (minimum warming rates to avoid "devitrification", or significant ice formation during warming from a vitrified state) are typically two or more orders of magnitude greater than critical cooling rates (Hopkins et al. 2012).

The same stringent and even contradictory conditions are imposed on storage temperature and transport of macroscopic samples, because masses of vitrifiable tissues larger than a few cubic centimeters almost invariably develop large-scale fractures. Storing them at liquid nitrogen temperature leads to cracking due to shear stress relaxation, while keeping them close to the glass transition temperature (down to 15 degrees) results in the formation of nanoscale ice crystals due to lateral diffusion of water molecules, which increases the critical warming rate upon subsequent heating.

A detailed review of the entire spectrum of research requires a separate report, and goes beyond the scope of this work. The more so because no major success has been achieved yet, excluding successful vitrification of blood vessels (1996), peripheral nerves, pancreatic islets and so on. Although speaking of nerves and blood vessels, we consider it to be rather a tissue engineering than cryobiology.

The most promising advances in this field were made by "21st Century Medicine", a company led by Gregory Fahy and Brian Wowk. In 2005 they have reported on a rabbit kidney that survived vitrification and subsequent transplantation with immediate contralateral nephrectomy, successfully functioning for 9 days (Fahy et al. 2009). However, due to the above-mentioned unresolved problems, this experiment remains anecdotal so far.

We know three possible ways of macroscopic biological objects accelerated warming: the dielectric warming (Wusteman et al. 2004), vascular perfusion with inert fluids that remain liquid at cryogenic temperatures (Federowicz, Harris et al. 1999), and gas perfusion (Schimmel et al. 1964; Bickis and Henderson 1966; Hamilton et al. 1973; Van Sickle and Jones 2014)

Dielectric warming results in very uneven heating.

Cooling solutions to cryogenic temperatures directly by cryoprotectants is impossible, due to progredient solute viscosity and peripheral vascular resistance elevation. As a result, the perfusion rate decreases. From a certain moment, instead of passing through the microvasculature, refrigerants begin to "shunt" through the major vascular arcades. Respectively the heat exchange is complementary weakened and cooling becomes irregular. With a further increase viscosity perfusion either stops or ruptures in the parenchyma occur. Liquid perfluorocarbon-based coolants are also unable to solve the problem of sufficient cooling rate for the same reasons, albeit at lower temperatures.

Gaseous coolants, despite their negligible heat capacity, have two to three orders of magnitude lower viscosity, and freely pass through the vascular bed even at cryogenic temperatures. The most popular in this area is helium, due to its low condensing temperature and the highest diffusion mobility.

Of course, the gas perfusion has its own shortcomings:
- low heat capacity forces the use of high pressure to increase the heat transfer, which is fraught with barotrauma;
- gas embolism—particularly during rewarming, when the gaseous coolant is replaced by a liquid;
- "drying up" of endothelium and the adjacent tissue areas.

Discussion and project elaboration

In view of the foregoing, for vitrification of isolated organs (and eventually up to intact organisms), suitable for transplantation after storage and rewarming, we plan the following:
1. Application of combined use of traditional cryoprotectants with clathrate forming gases as a means for vitrification. We turned

our attention to xenon (Prehoda 1969; Rodin et al. 1984; Shcherbakov, Tel'pukhov et al. 2004; Sheleg et al. 2008) because of its absolute nontoxicity on tissue and cellular level, while even the most low-toxic cryoprotectants (such as glycerol, ethylene glycol etc.) are potentially dangerous due to osmotic effects.

In order to evaluate the cryoprotective potential of xenon we have conducted two series of experiments—with baker's yeast (S. cerevisiae) and mammalian cell cultures (CHO-K1 and NIH-3T3) in custom-built bronze miniature hyperbaric chambers.

Yeast survival at all xenon pressures tested (3 to 7 at) after 2 hours of pressure chamber exposure at $-20°C$ with subsequent chamber immersion in liquid nitrogen (68,5 plus or minus 6%) turned out to be better than in control (35 plus or minus 6%) and comparable or even higher to that of 5% DMSO or glycerol application (50 plus or minus 25%). Joint application of xenon + DMSO or glycerol showed 71 plus or minus 15% of cell survival.

However, experiments with mammalian cells at xenon partial pressures of 2.5 to 12 at upon slow programmable freezing (SPF) showed complete cells dissolution upon conventional rewarming. Even after designing a sophisticated decompression protocol we could not achieve survival upon thawing. This made us disappointedly conclude that xenon has no cryoprotective properties applying SPF. Nevertheless, 2,8 plus or minus 2,3% of control group cells frozen by the abovementioned "yeast" protocol of cells survived and, moreover, fastest achievable cooling speed (direct placing of pressurized hyperbaric chamber into liquid nitrogen, which we considered an absolute death sentence) in the other control experiment improved cells survival up to 22,5 plus or minus 13,4%.

Large scale molecular dynamics simulations of xenon clathrate hydrate and ice crystal growth (details are recently described in our report "Possible Mechanisms of Cryoprotective Effect of Xenon") indicate vitrification of water inside cells, especially under combined application of xenon and DMSO, as a possible explanation of these results.

The main thing for us is that the hydrophobic solubility of xenon in low temperature water increases with cooling faster than the thermal stability of clathrate hydrate (Artyukhov et al. 2014).

As a result, it should turn out that traditional cryoprotectants, by suppressing xenon crystalline hydrate formation on cooling stage, would allow the solution to be saturated by xenon to a significant increase in liquid viscosity. This will enable vitrification using a much lesser amount of chemical cryoprotectants, and thus yield large decrease in toxicity. We have submitted a patent application.

Sounds so simple, even primitive. Why didn't anybody think of this before?

2. Second critical step is to develop a methodology for a fundamentally new approach to a uniform heating of complex biological objects by means of electric and magnetic fields' phased emitters, guided by MRI thermometry—also patent pending and discussed in yet another corresponding report at this conference.

3. Creation of an improved carrier solution for cryoprotectants with an emphasis on organ conditioning, prevention of chilling injury, neutralization of reactive oxygen species, activation of apoptosis cascades and overall reperfusion syndrome during rewarming.

4. Slight improvement of gas perfusion methodology. Increasing the heat exchange efficiency should be achieved through a conjunction of elevated persufflation pressure with modulation of sonic and ultrasonic waves.

5. Application of frequency and width-pulse electromagnetic field modulation within the initial stages of cooling, not confining to a simple 50 Hertz oscillation used in Cells Alive System low-temperature freezer of Abiko corporation (Owada and Saito 2010).

6. Determination of the best storage temperature for objects, vitrified by our method of (taking into account the alleged stabilizing effect of gas cryoprotectants).

References

Armitage WJ (1991) Preservation of viable tissues for transplantation. In: Fuller BJ, Grout BWW (Eds.). Clinical Applications of Cryobiology. CRC Press, Boca Raton 169–190

Artyukhov VI, Pulver AY., Peregudov A, Artyuhov I (2014) Can xenon in water inhibit ice growth? Molecular dynamics of phase transitions in water-Xe system. J Chem Phys 141: 034503. doi:10.1063/1.4887069

Fahy GM, Wowk B, Pagotan R et al. Physical and biological aspects of renal vitrification. Organogenesis. 5: 167–75

Federowicz MG, Harris S B et al. (1999) A method for rapid cooling and warming of biological materials. United States Patent 6274,303, 1999

Bickis IJ and Henderson IWD (1966) Helium Perfusion Technique Advances. JAMA 197: 36–7

Hamilton R, Holst HI et al. (1973) Successful preservation of canine small intestine by freezing. J Surg Res 14, 313–8

Hopkins JB, Badeau R, Warkentin M, Thorne RE (2012) Effect of common cryoprotectants on critical warming rates and ice formation in aqueous solutions // Cryobiology 65: 169–78

Isachenko V, Alabart JL, Dattena M et al. (2003) New technology for vitrification and field (microscope-free) thawing and transfer of the small ruminant embryos. Theriogenology 59: 1209–18

Luyet BJ (1937) Differential staining for living and dead cells. Science 85: 106

Mazur P (1984) Freezing of living cells: mechanisms and implications. Am. Physiol 247: C125–C142

Mazur P A (2010) Biologist's view of the relevance of thermodynamics and physical chemistry to cryobiology. Cryobiology 6: 4–10

Owada N, Saito S (2010) Quick Freezing Apparatus and Quick Freezing Method. United States Patent 7810, 340 (2010)

Polge C, Smith AU. Parkes A S (1949) Revival of spermatozoa after vitrification and dehydration at low temperatures: Nature 164. 666

Rall WF., Fahy GM (1985) Ice-free cryopreservation of mouse embryos at –196 degrees C by vitrification. Nature 313: 573–5

Prehoda RW (1969) Suspended Animation: The Research Possibility that May Allow Man to Conquer the Limiting Chains of Time. Chilton Book Company, Southborough

Rodin VV, Isangalin FS et al. (1984) Structure of protein solutions in a presence of xenon clathrate. Cryobiology & Cryo-Medicine 14. 3–7

Shcherbakov PV, Tel'pukhov VI, Nikolaev AV (2004) Method for Organs and Tissues Cryoconservation In Situ. RU patent 2268590 C1 (15 June 2004)

Sheleg S, Hixon H; Cohen B et al. (2008) Cardiac Mitochondrial Membrane Stability after Deep Hypothermia using a Xenon Clathrate Cryostasis Protocol—an Electron Microscopy Study. Int J Clin Exp Pathol 1: 440–7

Schimmel H, Wajcner G, Chatelain C et al. (1964) Freezing of whole rat and dog kidney by perfusion of liquid nitrogen through the renal artery. Cryobiology 1: 171–5

Van Sickle S (2014) Method and apparatus for prevention of thermomechanical fracturing in vitrified tissue using rapid cooling and warming by persufflation. United States Patent 5952,168, 2014

Wowk B (2010) Thermodynamic aspects of vitrification. Cryobiology 60: 11–22

Wusteman M, Robinson ML, Pegg D (2014) Vitrification of large tissues with dielectric warming: biological problems and some approaches to their solution. Cryobiology 48: 179–89

Uniform Heating of Multi-Structural Biological Objects by Means of Electric and Magnetic Fields' Phased Emitters

N. Shamaev, A. Pulver, D. Buslov

Abstract

We describe an approach to achieving uniform heating of bulky multi-structural biological objects by coherent electromagnetic radiation, guided by the data from the MRI thermometry. The explanation of how we plan to "focus" electromagnetic fields which in principle cannot be focused, is also given. We hope that this technique (patent pending) will allow to reach previously unattainable speeds of rewarming from cryogenic to near-zero temperatures, including non-vascular areas and cavities. This will allow achievement of rapid and uniform warming rates that are necessary to avoid devitrification even with reduced cryoprotectants concentrations, thus eliminating the main problems in applying vitrifying techniques to bulky tissues and organs

The main problems in applying vitrifying techniques to bulky tissues and organs are related to the difficulty in achieving sufficiently rapid and uniform warming rates that are necessary to avoid devitrification. Non-uniformity of temperature increases the risk of mechanical stresses and fractures developing in the glass during rapid warming.

Heating by conduction becomes increasingly inadequate as the size of the tissue increases and, once the tissue exceeds a few millimeters in thickness, can lead to excessively high temperatures on the periphery when the core remains vitrified. Such large temperature gradients also lead to mechanical stresses, fracture and tissue damage.

Electromagnetic absorption (dielectric heating) as an alternative method for warming of bulky tissues has been the subject of study for over 25 years. Early studies on whole tissue using conventional microwave ovens operating at frequency of 2450MHz were not successful. It was later established that a lower frequency would give a greater penetration depth and better uniformity of heating, and that the optimum frequency range for dielectric heating of tissues was 300–1000MHz. Uniformity of warming in a dielectric field is heavily dependent on the dielectric properties of the aqueous phase and, in particular, its cryoprotectant component.

Still, the best result was a purpose built power source and resonant applicator operating at 434MHz was that produced relatively uniform heating across 36mm gelatin spheres containing various concentrations of dimethylsulphoxide, with the final maximum temperature difference not exceeding 9°C [Wusterman et al., 2004].

We describe an approach to achieving uniform heating of bulky multi-structural biological objects by coherent electromagnetic radiation with zero field intensity region, guided by the data from the MRI thermometry.

Our method consists in the use of high-frequency heating in order to uniformly heat of heterogeneous tissues. For that we suggest to use a proposed focusing of coherent electromagnetic emission to obtain regions with zero electric and magnetic field component.

Creating a point of zero field strength will be obtained by placing a minimum of four emitters in the corners of a tetrahedron. The power of emission and the phase of emitters is defined on the basis of the provisions of the "zero point" within it.

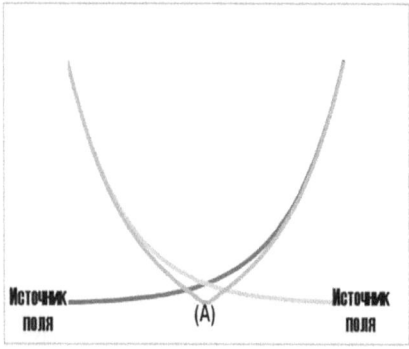

Fig 1.
Two-dimensional interference pattern of emission from two emitters.

Fig. 1 shows the field strength of the two emitters which are functioning with a shift of 180 °. The total field strength is shown in black. At the point A, the field intensity is zero.

We're changing the emission power of the emitter. So the "point" (A) moves on the X-axis between the emitters.

We solved the system of linear equations in which the initial data are:
- size of the object;
- the heat capacity of each virtual object point (VOP);
- absorptive capacity in each of the VOP;
- the initial value of the temperature in each of the VOP;

The solution is represented by the matrix with the following set of elements:

№	Coordinates	Power of emitters 1	Power of emitters 2	Phase of emitters 1	Phase of emitters 2	Time of heating
1						
2						
3						
4...						

The matrix will determine the work of the magnetic and electric fields of the emitters. The application of electric field is more complicated due to nature of its distribution.

There is a reason why we suggest to use two types of fields. Objects at low temperatures are more effectively heat by the capacitive method. A functioning of magnetic field depends from an electrical conductivity of the heated material. At the same time, objects with higher temperatures are more effectively heat by magnetic field.

Application of this method requires a solving of the matrix for at least four emitters. More sources of emission such as inductors for the magnetic field and the capacitors for electrical is associated with difficulties of solving the matrix. So it would be helpful to state that the emitters are located on the surface of the conceptual sphere in order to reducing of the dimensionality of the system. More precisely, the set of nested spheres. So we could avoid excessive tension on the borders of the field heated object.

Basically, we calculate phases, capacities and time of concentrate of the point (A) in each VOP ("concentrate non-heat"). It could be done by using a matrix obtained by a three-dimensional MRI thermometer [Basgal 2008]. Raw date is getting through information processing of variations, as result we have "three-dimensional matrix of specific heats".

Most basic application of the method could be represent as move of virtual point of non-heat with high speed in volume of the heated sample. It's like deflected beam in a television picture tube. More accurate likeness could be represented as accelerated in the hundreds of thousands times repetitive work of 3D-printers.

We could use more effective patterns of "movement" depending on the structure of the heated object. But they have to be discovered first.

Mathematical tools are extremely difficult in any case. We have to use either supercomputers, or create a hardware implementation of the math module using a large number of FPGAs in order to monitor the temperature of the object in real-time.

We hope that this technique (patent pending) will allow to reach previously unattainable speeds of rewarming from cryogenic to near-zero temperatures, including non-vascular areas and cavities. This will allow achievement of rapid and uniform warming rates that are necessary to avoid devitrification even with reduced cryoprotectants concentrations, thus eliminating the main problems in applying vitrifying techniques to bulky tissues and organs.

References

Duke BM (2008) Innovations improve accuracy of MRI as internal "thermometer" Duke Today October, 2008. http://shar.es/RHQW1.
Wusteman M, Robinson M, Pegg D (2004) Vitrification of large tissues with dielectric warming: biological problems and some approaches to their solution. Cryobiology 48: 179–89

Trying to Outwit Nature

Peter Gouras

Abstract

I have tried to show how my own lab has pursued repair of the central nervous system, the retinal epithelium and the photoreceptors. We still have a long way to go but some success is already apparent. To go further one needs optimism, imagination and experimentation. We may have the ability to make the blind see and the demented think logically again.

Introduction

Trying to outwit Nature is not an easy matter. She discovered a powerful strategy called Darwinian evolution, presumably by trial and error and a quasi-infinite amount of time that has led to the evolution of the most powerful machine in our universe, the human brain. In her scheme there is no room for immortality, only reproductive success, after which death clears the deck for further competition. This testing goes on as the world population expands following the dictum of Darwinism but not politicians that "all men are created unequal".

Overview of own results and conclusions

We comprise a few biological sports that believe we can conquer death, at least brain death. Are we wasting our time? I am not sure but it is an intriguing problem to tackle and think about. Here is our current strategy. We must cool the brain as soon as possible after clinical death, a difficult variable to control. Then we must perfuse the body and brain with an optimal cryo-protective; what we have now is not perfect. We must then cool the body and brain to liquid nitrogen temperatures without damaging these structures and store them for an indefinite time before attempting to rewarm and revivify them. Lastly we have to correct the problem that killed the subject and if possible rejuvenate the brain. These are very difficult tasks that may never be solvable. Experimentation is the only way to try to solve them. Several labs are trying to do this in the USA. Others have recently appeared in Germany, England and Russia. This idea is not going away.

My own research bears on some of these topics. In particular repairing thee central nervous system. I have been studying how the retinal epithelium ages and how it can be repaired. The retinal epithelium is a thin monolayer that forms the blood brain barrier of the neural retina and is part of the central nervous system. This epithelium nourishes the photoreceptors and if it fails, the photoreceptors degenerate and blindness ensues. This epithelium does not divide and the layer you are born with remains with you until you die. One of its nourishing tasks is to phagocytose the growing, lipid tips of the photoreceptors and dissolve them. With time its efficiency for doing this declines and there is a buildup of lipid filled lysosomes which begins to choke and kill the epithelium leading to blindness. In fact this is the leading cause of blindness in humans over the age of fifty. Why this epithelium loses its ability to phagocytose and dissolve the ingested outer segments of photoreceptors is a key question in the pathogenesis of this disease. I favor the hypothesis that it is due to oxidative stress emanating from mitochondrial production of reactive oxygen species. This is one of the leading hypotheses for the cause of aging in general. I favor this explanation of why the retinal epithelium becomes faulty in its lysosomal degradation of lipids. It is based on my comparison of the same problem in man and monkey. In the monkey the dysfunction of the retinal epithelium occurs much more rapidly with age than it does in man. This suggests to me that the monkey epithelium is less able to handle oxidative stress than human retinal epithelium. This is supported by the fact that monkey ages much more rapidly than man.

But how can we repair this problem of aging retinal epithelium dysfunction? Several years ago I demonstrated that one could surgically remove this epithelial monolayer and replace it with cultured retinal epithelium. Together with surgeons at the Karolinska Institute in Sweden we tried to transplant embryonic retinal epithelium into human subjects with a late stage destructive phase of age related macular degeneration. This attempt failed for several reasons. One was a surgical technique that did not allow us to directly see the epithelial layer and reestablish a new layer. The second problem was that we were transplanting foreign cells which faced host/graft rejection. Now these two problems can be corrected. A new surgical technique allows for direct visualization of the degenerate epithelial layer and the use of the subject's own cells for transplantation should eliminate rejection. It has now become possible to biopsy skin fibroblasts and convert them into pluri-potential cells in culture and then convert the pluri-potential cells into retinal epithelium. I bring this up to demonstrate how progress can be made in repairing a unique part of the central nervous system, the retinal epithelial layer. It took years of experimentation.

To do this for brain cells, like neurons and glia, is much more difficult but not inconceivable. We have experimented with photoreceptor transplantation. Photoreceptors are very active neurons but they have one great advantage for transplantation; only their central output end must make synaptic contact with second order neurons. For most neurons in the brain both their input and output require specific synaptic connectivity, which is more difficult. We have shown that one can transplant embryonic murine photoreceptors into another mouse that has lost its photoreceptors due to a genetic defect. There is excellent survival of the transplants if they are properly oriented to the retinal epithelium. If they are not they degenerate. The properly oriented transplants grow light sensitive outer segments and survive indefinitely. What we have not been able to demonstrate unequivocally is the formation of synapses of these transplants on second order neurons. It requires more experimentation that examines how neurons are activated appropriately to form synapses with their proper targets. This will certainly come with more time.

References

All references to this research can be obtained from Pub Med using Gouras P

New Advances in Organ Cryopreservation. Electromagnetic Rewarming and Selective Targeting of Ice Nuclei

Ramon Risco, Alberto Olmo, and Pablo Barroso

Abstract

In this chapter we will review the most recent technologies suggested and tested for improving organ rewarming and we will propose a new technique, based on the combination of other previous works, which we think can hold the key to the final solution of the cryopreservation of organs.

1. Introduction: The problem of cryopreservation of organs

Hundreds of people die every day waiting for an organ. The demand for organ transplants is growing at a rate five times higher than the availability of the same, and the urgency of the transplant is the fundamental reason for the lack of matching and loss of organs. The ability to create engineered tissue organs from stem cells has also generated an urgent need to store these organs. State-of-the-art techniques for cryopreserving bulk biomaterials and organ systems will transform in the near future current approaches in different fields, such as transplantation, regenerative medicine or pharmacology.

In recent years there have been important results achieved in the cryopreservation of biological material. Different strategies have been developed to prevent and control the formation of ice during the process of cryopreservation of cells and tissues, such as vitrification, with excellent res (Mc Intyre and Fahy 2015; Lewis et al. 2186). Important advances have been reported with the cryopreservation of several animal organs, such as rabbit kidneys (Lewis et al. 2106) and sheep ovaries (Campbell et al. 2014).

However, the cryopreservation of human organs has been a challenge that we have not managed to overcome so far mainly due to the difficulty in eliminating the formation of ice inside the organ (Lewis et al. 2016; Corral et al. 2015). This of special relevance in the rewarming process, as the temperature increase rates required to avoid significant ice growth during this phase are larger than the cooling rates required during the prior vitrification, when the sample is relatively un-nucleated (Wowk 2015) The fast and uniform rewarming of the organs from the vitrified state is one of the main challenges in modern cryopreservation.

The aim of this paper is the improvement of organ rewarming. We will propose a new technique, based on the combination of other previous works,

2. Advances in electromagnetic rewarming

A lower frequency for the electromagnetic fields has been proposed by B. Wowk et al. (Wowk, Corral 2013). According to this work, the temperature zone where previously-nucleated ice crystals grow most rapidly typically spans tens of degrees Celsius below the melting temperature, above which ice cannot exist, and below which ice growth is kinetically inhibited by viscosity (Wowk, Corral 2013). The frequency used for electromagnetic warming should be chosen to couple well to the cryoprotectant solution in this temperature zone, while coupling poorly at higher temperatures. Such a frequency will cause rapid passage through the ice growth temperature zone and then cause warm spots to loiter above the melting temperature while colder parts of the organ catch up, thereby avoiding thermal runaway. According to initial experiments carried out for the M22 vitrification solution that has been used for rabbit kidney vitrification in their laboratory, the temperature zone of rapid ice growth is approximately −75°C to −60°C. The solution viscosity in this temperature zone is very high, with a Debye dipole relaxation frequency on the order of 10MHz. This suggested an electromagnetic warming frequency much lower than the hundreds of megahertz typically studied in cryobiology (Wowk, Corral 2013).

However, practical application still needs further improvements, due to diffusive heat and mass transfer limitations, which are typically manifested as devitrification and cracking failures during rewarming.

2.1. Magnetic nanoparticles (mNP)

Magnetically heated nanoparticles have also been recently tested to improve the electromagnetic rewarming process, improving uniformity and rapidity in the process (Etheridge et al. 2014; Wang and Zhao 2016). The excitation fields used (alternating magnetic fields at hundreds of kHz) are relatively transparent to biological tissues, but generate significant heating in magnetic particles, which can be distributed throughout macro- and microscopic tissue structures. The homogeneity in heating then depends on the magnetic particles distribution and the frequency of the electromagnetic fields used. One major advantage reported is also the tunability of the heating rate, where the applied field can be adjusted to better control rewarming protocols (e.g. annealing to reduce thermal stresses around the glass transition (Etheridge and Xu 2014).

In other similar work (Wang et al. 2016), the successful synthesis and application of Fe_3O_4 nanoparticles for magnetic induction heating to enhance rewarming of vitrification-cryopreserved human umbilical cord matrix mesenchymal stem cells (hUCM-MSCs) is reported. The enhanced rewarming with magnetic nanoparticles greatly improves the survival of vitrification-cryopreserved hUCM-MSCs. Moreover, the hUCM-MSCs retain their intact stemness and multilineage potential of differentiation post cryopreservation by vitrification with the enhanced rewarming.

Again, the final application of this technology to achieve the effective recovery of organs, avoiding devitrification and cracking failures in bulk biological material, as reported in Ethridge et al. (2014); Wang, et al. (2016) would require further improvements and the combination of several techniques. In the following section we will describe our proposal, based on the combination of magnetic nanoparticles and ice-binding proteins.

3. New ideas for organ cryopreservation: Selective targeting of ice nuclei before rewarming

The effective elimination of the small ice nuclei existing in the vitrified sample before the rewarming process would enormously facilitate the results obtained, either if the rewarming is carried out by traditional methods or if it is carried out by other methods such as in the presence of electromagnetic fields (Lewis et al. 2016).

An initial experimental evidence of the benefits of the elimination of ice prior to rewarming can be found in a work by Fowler and Toner (1998), where a laser pulse was used to selectively target and melt the intracellular ice existing in erythrocytes, to then achieve the successful recovery of the cells.

The present proposed idea aims to sort out this effective elimination of small nuclei of ice in bulky biological material and whole organ systems before the beginning of the rewarming process, selectively targeting ice nuclei inside the sample, and therefore establishing the basis for the correct recovery procedure of the organ.

The suggested approach is mainly based on the union of magnetic nanoparticles (mNP) and ice-binding proteins (IBP), which have been already addressed in a separate way by different groups in the past. In sub-section 3.1 we will briefly introduce ice-binding proteins, to finally describe our proposal in sub-section 3.2.

3.1. Ice-binding proteins (IBP)

One of the strategies organisms have developed to thrive in cold ecosystems where there is a risk of freezing is the production of ice-binding proteins (IBPs). As implied by their name, IBPS bind to ice surfaces and in doing so control ice growth to help organisms coexist with ice and avoid freeze injury. Since the discovery of antifreeze glycoproteins (AFGPs) in the late 1960s in the blood of Antarctic notothenioid fishes (DeVries et al. (1970); DeVries and Wohlschlag 1969) various types of IBPs have been identified in many biological kingdoms (Doley et al. 2016). Important ice adhesion functionalities have also been recently discovered by

Guo et al. in bacterial antifreeze proteins (Guo et al. 2012). Even though with the only presence of IBPs organ cryopreservation cannot be fully achieved, IBP may hold the key to it, in combination with other techniques, as shown in section 3.2.

3.2. mNP—IBP particles

Proteins have the possibility to attach magnetic nanoparticles, as 2shown in different works (Latigue et al. 2012). In that work, a method based on nanomagnetism that allows a specific in situ monitoring of interactions between iron oxide nanoparticles and blood plasma is proposed. Tracking the nanoparticle orientation through their optical birefringence signal induced by an external magnetic field provides a quantitative real-time detection of protein corona at the surface of nanoparticles and assesses eventual onset of particle aggregation. In principle, as shown in (Latigue et al. 2012), a mNP-IBP complex can be built; we propose to call it "magnetic ice-binding protein (mIBP)". When trying to vitrify an organ, if the vitrifying solution contains mIBPs, the mIBPs will attach to the small ice embryos that eventually can be formed during cooling to cryogenic temperatures.

After storage, but before the rewarming process, the idea is to produce the induction heating of the mIBP with a very short radiofrequency pulse, melting the ice embryo to which it is attached. The close proximity of the bulky vitrified material at cryogenic temperatures will cool fast enough the melted ice and will vitrify it too (Fowler and Toner 1998). The frequency to be used will be slightly different to the excitation fields used for magnetic nanoparticles reported in (Etheridge et al. 2014; Wang et al. 2016), usually at hundreds of kHz.

Once the different ice nuclei are eliminated (melted and vitrified by this mechanism), the rewarming process of the cryopreserved material can be carried out as proposed in other works, for example through electromagnetic rewarming (Wowk 2015; Wowk and Corral 2013)), sorting out the problems found with the presence of ice nuclei, and finally avoiding problems of devitrification and cracking in the bulk biological material.

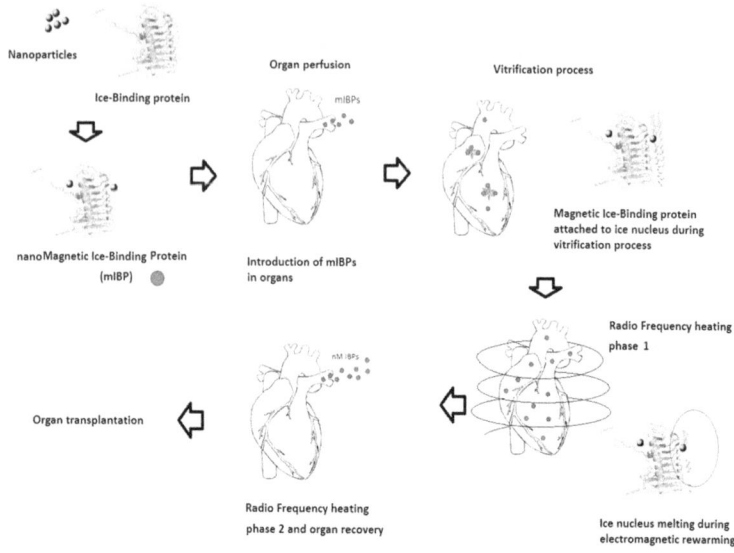

Fig 1.
Scheme of the proposed technique. Magnetic nanoparticles are attached to an Ice-Binding protein, forming a magnetic Ice-Binding Protein (mIBP). The new composed mIBP is attached to a small ice nucleus during vitrification. Before rewarming, when excited at the right frequency, the magnetic nanoparticle produces the ice nucleus heating and melting. Finally, the rewarming process is completed, in the absence of ice nuclei.

Work is ongoing at the University of Seville to examine the viability of building magnetic ice-binding proteins, optimizing the electromagnetic pulse used, and examining the efficiency of the process in the rewarming of organs. Figs. 2 and 3 show some of this initial work performed.

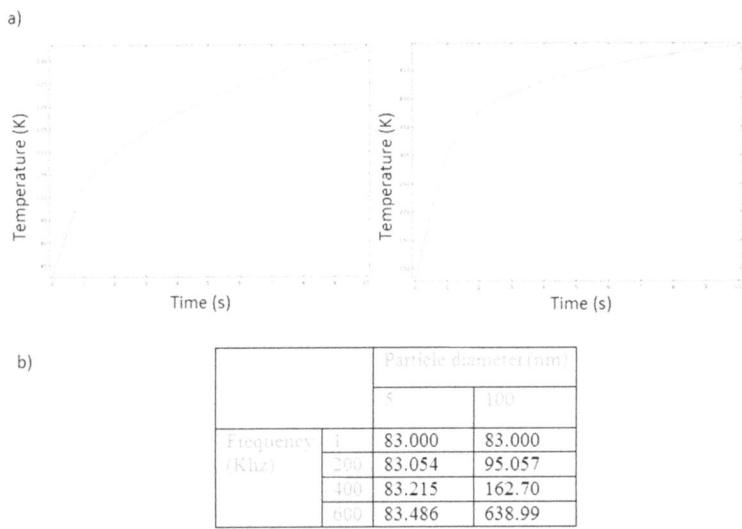

Fig. 2.
Simulations in COMSOL of the electromagnetic warming of nanomagnetic particles. a) Evolution of temperature in time of different magnetic particles in the presence of electromagnetic fields of different frequencies. Left: 200nm diameter at 200Khz. Right: 500nm diameter at 400Khz. b) Final temperatures achieved by nanomagnetic particles of 5 and 100 nm, starting at 83K, after 10 seconds of exposure to electromagnetic fields at different frequencies, with 12A intensity. The best results are obtained with 100 nm particles at a frequency of 600Khz, where the fastest rewarming is achieved. Simulations were performed with COMSOL Multiphysics.

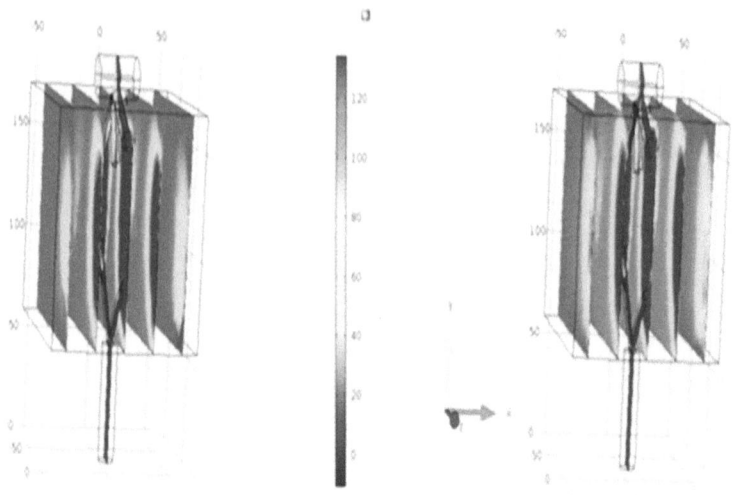

Fig. 3.
Simulation of the perfusion of organs with nanomagnetic particles. Simulation of the perfusion of organs (liver in this case) perfused with nanomagnetic particles in COMSOL Multiphysics. A modification of the model of a pork liver by Rohan et al. (2014) was used. Work is ongoing to select the right type of magnetic nanoparticles (material and size), the right concentration of mIBPs, and the right type of frequency used for the electromagnetic rewarming.

4. Conclusions

The cryopreservation of human organs has been a challenge that we have not managed to overcome so far due to the difficulty in eliminating the formation of ice inside the organ (Lewis et al. 2016). This is of special relevance in the rewarming process. Electromagnetic rewarming of vitrified tissue has proved to be superior to external conduction for rapidly and uniformly traversing the temperature zone of maximum ice growth where risk of devitrification is greatest. Magnetic nanoparticles have also been initially tested to improve the electromagnetic rewarming process, improving uniformity and rapidity in the process.

A final step to improve the process is presented, combining the use of ice-binding proteins with magnetic nanoparticles to melt the existing ice nuclei in a more efficient way at the beginning of the rewarming process. We think this combination will finally open the door to the unlimited storage of organs at cryogenic temperatures. Work is ongoing to optimize the technique, selecting the right type of magnetic particles, proteins, and the right frequency to be used in the electromagnetic warming.

References

Campbell BK, Hernandez-Medrano J, Onions W et al. (2014) Restoration of ovarian function and natural fertility following the cryopreservation and autotransplantation of whole adult sheep ovaries. Hum Reprod 29: 1749–63

Corral A, Balcerzyk M, Parrado-Gallego A, et al. (2015) Assessment of the cryoprotectant concentration inside a bulky organ for cryopreservation using X-ray computed tomography. Cryobiology 71: 419–31

DeVries AL, Komatsu SK, Feeney RE (1970) Purification and characterization of a freezing-point-depressant glycoprotein from Antarctic. J Biol Chem 245: 2901–8

DeVries AL, Wohlschlag DE (1996) Freezing resistance in some Antarctic fishes. Science 163: 1073–5

Dolev MB, Braslavsky I, Davies PL (2016) Ice-Binding Proteins and Their Function. Annu Revi Biochemi 85: 515–42

Etheridge ML, Xu Y, Rott L, et al. (2014) Heating of magnetic nanoparticles improves the thawing of cryopreserved biomaterials. Technology 2: 229

Fowler AJ, Toner M (1998) Prevention of Hemolysis in Rapidly Frozen Erythrocytes by Using a Laser Pulse. Ann N Y Acad Sci 858: 245–52

Guo SQ, Garnham CP, Whitney JC et al. (2012) Re-evaluation of a bacterial antifreeze protein as an adhesin with ice-binding activity. PLoS ONE 7

Latigue L, Wilhelm C, Servais J et al. (2012) Nanomagnetic Sensing of Blood Plasma Protein Interactions with Iron Oxide Nanoparticles: Impact on Macrophage Uptake. ACS Nano 6: 2665–78

Lewis JK, Bishof C, Braslavsky K et al. (2016) The Grand Challenges of Organ Banking: Proceedings from the first global summit on complex tissue. Cryobiology 72: 169–82

Marsland TP, Evans S, Pegg DE (1987) Dielectric measurements for the design of an electromagnetic rewarming system, Cryobiology 24: 311–23

McIntyre RL, Fahy GM (2015) Aldehyde-stabilized cryopreservation. Cryobiology 71: 448–58

Rohan E (2014) Towards Microstructure Based Tissue Perfusion Reconstruction From CT Using MultiScale Modeling. Blucher Mechanical Engineering Proceedings 18

Wang J, Zhao G, Zhang Z et al. (2016) Magnetic induction heating of superparamagnetic nanoarticles during rewarming augments the recovery of hUCM-MSCs cryopreserved by vitrification. Acta Biomater 33: 264-74

Wowk B (2015) Considerations for electromagnetic warming of vitrified biomaterials. Cryobiology 71: 176

Wowk B, Corral A (2013) Adaptation of a commercial diathermy machine for radiofrequency warming of vitrified organs. Cryobiology 67: 251-2

We thank financial support of Excellence Project 2008 of Junta de Andalucia.

Forms of Cryopreservation Damage and Strategies for Prevention or Mitigation

Ben Best

Abstract

The different forms of damage observed during cryopreservation are entitled and techniques of avoiding and preventing those damages are discussed.

Introduction

Cryopreservation of organs or human cryonics patients at cryogenic temperatures can be a way of stopping biological deterioration for a very long time. Although cryopreservation can be damaging, there are strategies that can be used to prevent or mitigate that damage. The nine forms of damage to be discussed in this article are: (Hays et al. 2001) chilling injury (Rojas et al, 1996) damage due to ice (Mazur et al. 1992) fracturing due to thermal stress (Fahy et al. 2004) nucleation at low temperature (Lemler et al. 2004) ice growth on rewarming (Pichugin et aol. 2006) cryogenic storage injury (Fahy et al. 2009) ischemic damage due to slow cooling or ice temperature storage (Franks 2003) perfusion difficulties (Wowk 2010) cryoprotectant toxicity.

Chilling Injury

Cooling of biological tissues below the temperatures at which they normally function can be damaging, even above freezing temperature and without ice formation. Injury simply due to cooling is referred to as *chilling injury*. There may be many causes of chilling injury, the mechanisms are poorly understood. One mechanism of chilling injury can be a liquid-to-gel change of phase by lipids in cell membranes.[1] This phase change is comparable to grease solidifying in a frying pan upon cooling. Solidification of lipids in cell membranes causes the cells to be leaky and less functional.

Chilling injury in houseflies has been shown to be due to free radical damage (Rojas et al. 1996). In 1992 fruit fly embryos were vitrified without chilling injury by cooling at 20,000°C per minute (Mazur et al. 1992). Such rapid cooling would not be feasible for large biological samples. Hypertonic solutions in the range of 1.2 to 1.5 times isotonicity, however, have been used to completely abolish chilling injury during cooling from 0°C to −22°C.[4]

Damage Due to Ice

Cooling below 0°C can lead to damage due to ice. At 0°C ice is less dense than water, which means that ice formed by cooling tissues below 0°C will crush those tissues as expanded solid ice forms from liquid water. Liquid to solid crystalline phase transition can be avoided by inducing *vitrification*. Just as silicon dioxide forms a glass (and is prevented from forming a crystal) by adding soda (Na_2O) and lime (CaO), water can be induced to form a glass upon cooling by adding cryoprotective agents, such as glycerol, ethylene glycol, or dimethyl sulfoxide. With vitrification, a liquid becomes increasingly syrupy during cooling, finally forming a solid that is not a crystal (so viscous that it cannot be structurally distinguished from a liquid on the molecular level).

Cryopreserved rabbit brains perfused with vitrification solution have been cooled to −130°C without ice formation (Lemler et al. 2004). Vitrified hippocampal slices have been cooled to −130°C and rewarmed with complete viability (Pichugin et al. 2006). A vitrified rabbit kidney was cooled to −135°C, rewarmed, and transplanted back into the rabbit, allowing the rabbit to survive using the formerly vitrified kidney as its sole functioning kidney (Fahy et al. 2009). The kidney was, however, somewhat damaged. Creatine (a biomarker of kidney function) blood levels were excessively high.

Fracturing Due to Thermal Stress

The vitrification solutions M22 and VM-1 will solidify at about −124°C. This solidification temperature is called the *glass transition temperature*, represented by the symbol T_g. Further cooling of a vitrified sample from Tg down to −196°C (liquid nitrogen temperature) can cause cracking/fracturing for large biological objects due to thermal stress. When a solid with poor thermal conductivity is cooled externally, the outside of the sample contracts more (because it is colder) than the inside (which is warmer). This differential contraction results in thermal stress, which can cause the solid to crack or fracture.

Dr. Brian Wowk at 21 Century Medicine, Inc. has cooled a rabbit kidney from −124°C to −196°C over a two-day period without the kidney cracking. But a rabbit kidney is very small, only about 10mL. Much larger organs could take years, decades, or longer to cool to −196°C without cracking, which is generally not practical.

An Intermediate Storage Temperature (ITS) means storage just below −124°C, and well above −196°C. Cracking can be avoided by storage at ITS.

Nucleation at Low Temperature

For water to form ice, nuclei must be present. Absolutely pure water will not form ice above −40°C (Fahy et al. 2009), but silver iodide can form ice nuclei in clouds at −5°C to cause rain. Depending on the cryoprotectant and concentration of cryoprotectant, nuclei can form in vitrification solutions. Once nuclei have been formed, ice formation can begin, especially if the cryoprotectant concentration is low. The growth of crystals in a vitrified solution is called recrystallization. The rate of nucleation is highest at Tg (glass transition temperature), and ceases below −140°C. The rate of crystal growth is highest at −70°C (Wowk 2010). If nuclei have formed in a vitrification solution near Tg, there may be considerable ice growth at −70°C during rewarming. Vitrified pig brains can be cooled to −140°C without fracturing. Pig brains are much larger than rabbit kidneys, but are still much smaller than human brains. The ideal ITS for storing vitrified human brains would be above −140°C, but below −130°C, although the exact best temperature remains undetermined.

Ice Growth on Rewarming

Samples cryopreserved at ITS that have experienced ice nuclei formation face the danger of rapid ice growth at −70°C upon rewarming. Dr. Brian Wowk at 21st Century Medicine, Inc. has developed radiofrequency rewarming equipment that can rewarm at a rate of 200°C per minute at −70°C. Such fast warming should prevent ice growth. Faster warming rates are possible, but could cause damaging hotspots.

Cryogenic Storage Injury

Long-term storage at cryogenic temperature may avoid damage due to molecular motion, but does not necessarily protect against radiation damage. Acetylcholinesterase enzyme under X-ray irradiation shows conformational changes when irradiated at −118°C, but not at −173°C (Weik et al. 2001). subjected to X-ray irradiation showed four times as much damage to disulfide bridges when irradiated at −173°C compared to −233°C (Meentz et al. 2010). Such damage may not be a real cause for concern, however, if normal X-ray radiation does not match these experimental conditions, even over decades or centuries. Northern wood frogs spend months in the winter in a semi-frozen state at −3°C to −6°C with full recovery of heartbeat upon rewarming (Costanzo et al. 1995).

Ischemic Damage Due to Slow Cooling or Ice Temperature Storage

Ischemic damage begins when blood flow stops. Cooling will slow metabolic rate, and thereby slow ischemic damage. Between 40°C and 0°C reaction rates are reduced by one-third to one-half for every 10°C drop in temperature (Davidson and Janssens 2006). Cooling rate is increased and ischemic damage is reduced by using a slurry of ice water rather than ice cubes. Cooling rate is further increased by causing water to flow rapidly, carrying heat away from a cryonics patient.

Rather than use external cooling, a cryonics patient can be cooled more rapidly by perfusing a cold fluid through the blood vessels. But this form of cooling requires time-consuming surgery—time during which ischemic damage occurs. Liquid ventilation (cold fluid in the lungs) provides rapid internal cooling, and can be applied immediately without the need for surgical access. Good portable equipment for liquid ventilation is still under development by Suspended Animation, Inc.

Perfusion Difficulties

Cryonics patients become increasingly difficult to perfuse with vitrification solution if they have suffered too much ischemia, such as during long shipment times to a cryonics facility at ice-water temperature. Blood vessels become leaky or resist flow. Advanced Neural Biosciences (ANB) has developed vitrification solution additives which make the perfusate more viscous, and more capable of perfusing a moderately ischemic cryonics patient who would otherwise be difficult to perfuse. ANB wants more validation, however, before putting their perfusate into practice.

Ischemic damage due to shipment of cryonics patients in ice from Europe to the United States can be avoided by vitrifying the patient in Europe, and then shipping in dry ice. Although dry ice is warmer ($-78.5°C$) than Tg ($-124°C$), vitrified cryonics patients will not be subject to recrystallization over the few days required for shipment. This procedure was first applied to the wife of British cryonicist Alan Sinclair, who was vitrified in the UK and shipped in dry ice to the United States in May 2013.

An improved perfusion technique called Flow Lock has been developed by 21st Century Medicine, Inc. to prevent damage to a kidney being vitrified. Kidney blood flow is very unequal, with about 90% of blood going to the kidney cortex, about 10% going to the kidney outer medulla, and about 1–2% going to the kidney inner medulla. The challenge is to get adequate vitrification solution into the inner medulla without poisoning the cortex with excessive cryoprotectant. With Flow Lock perfusion, flow rate remains constant and pressure increases gradually as viscosity increases. (For more details on Flow Lock, see the video on this subject at www.societyforcryobiology.org 2013 annual meeting.)

Cryoprotectant Toxicity

Cryoprotectants are toxic, so cryobiologists attempt to use a concentration of cryoprotectant that minimizes toxicity while maximizing vitrification. If there was no concern about cryoprotectant toxicity, large amounts of cryoprotectant could be used, thereby eliminating nuclei formation near Tg or ice growth upon rewarming. Cryoprotectants become less toxic at lower temperature, which is why rapid cooling allows for vitrification solutions with higher cryoprotectant concentrations.

The most ambitious attempt at establishing a general theory of cryoprotectant toxicity is the qv* hypothesis of Dr. Gregory Fahy at 21st Century Medicine, Inc. "q" quantifies moles of water per mole of polar groups on a cryoprotectant molecule at the concentration needed to vitrify (v) under standardized conditions (*) (Fahy et al. 20004). "qv*" is proposed to measure the average hydrogen-bonding strength between cryoprotectant polar groups and water molecules in a solution. Plotting viability of kidney slices against qv* for a list of cryoprotectant solutions chosen by Fahy resulted in a declining straight line which Fahy suggested means that as the vitrifying power of cryoprotectant solution increases, the toxicity increases (Fahy et al. 20004; Fahy 2010). Fahy posits that cryoprotectants that strongly bind water will reduce the amount of water molecules available to protectively hydrate macromolecules, resulting in molecular damage.

Although the qv* hypothesis may be correct, a number of questions have been raised about the theory, such as the fact that it does not explain why mixtures of cryoprotectant chemicals can be less toxic, but more vitrifying (Best 2015).

Cryoprotectant toxicity is the most challenging and important technical problem in cryobiology and cryonics. Reversible organ cryopreservation at cryogenic temperatures could be easily achieved if less toxic cryoprotectant solutions were available. The mechanisms of cryoprotectant toxicity are not understood. The nature of cryoprotectant toxicity might become apparent by applying cryoprotectants and mixtures of cryoprotectants to biological samples and studying exactly what changes occur on a molecular level (Best 2015).

Success in such research could lead to the greatest breakthroughs in cryonics and cryobiology.

References

Best BP (2015) Cryoprotectant Toxicity: Facts, Issues, and Questions. Rejuvenation Res 18: 422–36

Costanzo JP, Lee RE Jr, DeVries AL et al. (1995) Survival mechanisms of vertebrate ectotherms at subfreezing temperatures: applications in cryomedicine. FASEB J 9: 351–8

Davidson EA, Janssens IA (2006) Temperature sensitivity of soil carbon decomposition and feedbacks to climate change. Nature 440: 165–73

Fahy GM (2010) Cryoprotectant toxicity neutralization. Cryobiology 60 (3 Suppl): 45–53

Fahy GM, Wowk B, Pagotan R et al. (2009) Physical and biological aspects of renal vitrification. Organogenesis 5: 167–75

Fahy GM, Wowk B, Wu J, Phan J et al., Cryopreservation of organs by vitrification: perspectives and recent advances. Cryobiology 48: 157–78

Fahy GM, Wowk B, Wu J, Paynter S (2004) Improved vitrification solutions based on the predictability of vitrification solution toxicity. Cryobiology 48: 22–35

Franks F. (2003) Nucleation of ice and its management in ecosystems. Philos Trans A Math Phys Eng Sci 361: 557–74

Hays LM, Crowe JH, Wolkers W, Rudenko S (2001) Factors affecting leakage of trapped solutes from phospholipid vesicles during thermotropic phase transitions. Cryobiology 42: 88–102

Lemler J, Harris SB, Platt C, Huffman TM. (2004) The arrest of biological time as a bridge to engineered negligible senescence. Ann N Y Acad Sci 1019: 559–63

Mazur P, Cole KW, Hall JW (1992) Cryobiological preservation of Drosophila embryos. Science 258: 1932–5

Meents A, Gutmann S, Wagner A, Schulze-Briese C (2010) Origin and temperature dependence of radiation damage in biological samples at cryogenic temperatures. Proc Natl Acad Sci U S A 107: 1094–9

Pichugin Y, Fahy GM, Morin R. 2006) Cryopreservation of rat hippocampal slices by vitrification. Cryobiology 52:228–40

Rojas RR, Leopold RA. (1996) Chilling Injury in the Housefly: Evidence for the Role of Oxidative Stress between Pupariation and Emergence. Cryobiology 33: 447–58

Weik M, Ravelli RB, Silman I, et al. (2001) Specific protein dynamics near the solvent glass transition assayed by radiation-induced structural changes. Protein Sci 10: 1953–61

Wowk B. (2010) Thermodynamic aspects of vitrification. Cryobiology 60: 11–22

Identification, Validation, and Implementation of New Cryonics Technologies (an essay)

Aschwin de Wolf

Abstract

In an ideal world, promising cryonics technologies would be identified, followed by prompt validation and implementation. In the real world, however, there are multiple reasons why potential improvements in cryonics are not being recognized or endorsed. Even when the benefits of such technologies appear evident, institutional and financial obstacles can prevent timely experimental validation and introduction. This article briefly reviews the history of technological progress in cryonics, discusses the reasons that delayed or postponed the introduction of superior technologies, and offers solutions that may enable faster adoption of new advances.

Introduction

The practical production of liquid nitrogen from liquefied air was first achieved by Carl von Linde in 1905, although liquid nitrogen only became widely available commercially after World War II. The idea of cryonics was introduced to the general public in the mid-1960s. Since liquid nitrogen (or liquid helium) is an essential requirement for human cryopreservation it is interesting to recognize that there was only a difference of roughly 20 years between cryonics being technically possible and the first efforts to practice cryonics. Robert Ettinger published *The Prospect of Immortality* in 1964. In 1967 James Bedford was cryopreserved.

Similarly, the idea of vitrification by rapid cooling as a means of cryopreservation was first proposed by Basil L. Luyet in the 1930s, followed by Pierre Boutron's screening of cryoprotectants for their glass forming abilities in the 1970s, and Gregory Fahy's pioneering work in the 1980s and beyond to achieve vitrification by high concentrations of cryoprotectants. No more than

20 years after these investigations, vitrification solutions with high concentrations of cryoprotectants were introduced in cryonics. This appears to be a reasonably rapid translation of scientific breakthroughs into cryonics technologies.

In the case of combinational pharmacotherapy to mitigate cerebral ischemia, research and cryonics implementation often went hand-in-hand and observations in cryonics cases were used to refine experimental designs.

Despite all this, there is the public perception that cryonics suffers from a lack of research and sees little technological progress. Compared to fields such as biogerontology and the developments discussed above, I think this is a misunderstanding. A major reason for it is that the general public and most scientists do not recognize that technological progress is possible in cryonics without creating full fledged human suspended animation. For example, safe and cost-effective cryogenic storage, inhibition of ice formation, elimination of (cerebral) ischemia, et cetera, are possible without having fully reversible cryopreservation.

I do think, however, that there is a lot that can be done to further narrow the time between identification, validation, and implementation of cryonics technologies by obtaining a greater understanding of what fosters and limits the identification of technological improvements in cryonics.

Identification of new technologies

Identification of new cryonics technologies is a topic that is rarely discussed within cryonics. Upon closer scrutiny, this is a rather complex topic. First of all, for the idea of identification of new technologies to make sense one has to subscribe to the idea that cryonics technologies can and must be improved. Closely related to this is the belief that the concept of "patient care" is meaningful in cryonics and can be empirically defined. This outlook on cryonics has not been universal and from its inception proponents of perfecting cryonics technologies often had to compete with a movement in cryonics that showed little interest in delivering cryonics services that aimed for more than placing the patient in liquid nitrogen after pronouncement of legal death.

The history of the Alcor Life Extension Foundation shows a different perspective. Since its inception, the organization has been shaped by individuals who aimed to close the gap between crude freezing and reversible human cryopreservation. One claim that I will be making in this article is that formal commitment to develop human suspended animation provides a framework to identify desirable research and development goals. When suspended animation is used as a benchmark to evaluate the state of cryonics technologies, it is possible to identify the gap between contemporary technologies and desired technologies. This, in turn, can direct the search for new developments in science and technology to replace existing technologies. For example, ice formation is clearly not compatible with human suspended animation and replacing freezing protocols with protocols that eliminate ice formation is a logical consequence of this mandate. Another example is fracturing. Long-term care protocols that induce too much thermal stress in the patient do not allow for reversible cryopreservation and need to be replaced with long term cryostasis protocols that avoid the formation of fractures, such as annealing or intermediate temperature storage (ITS).

It is important to stress here that a universal consensus to use human suspended animation as the ideal to strive for does not exclude debate over which new developments should be pursued and prioritized. I think there is a rather widespread consensus that the replacement of conventional cryopreservation with vitrification is highly desirable. But there can be a difference of opinion about how much effort to expend in developing completely non-toxic vitrification agents instead of accepting a small amount of toxicity and moving on to eliminating fracturing or cerebral dehydration first. Sometimes such differences in perspective reflect incomplete knowledge. For example, do we need to induce hypothermia faster during stabilization procedures, or are our existing technologies sufficient to keep the brain viable by contemporary medical criteria?

To my knowledge, no one in cryonics has ever attempted to offer a framework to make such decisions. In principle, such a framework should be possible. One could argue that the first mandate of a cryonics organization is to pursue technologies that preserve ultrastructure in such a state that no differences between controls and experimental brains can be observed. When this goal has been achieved, the next mandate is to eliminate gross mechanical damage, that is to say, prevent fracturing. The next step would be to prevent nano-scale modifications in proteins that compromise viability, that is, to develop non-toxic cryoprotectants. Such a ranking can also assist in cost-benefit analysis of proposed technologies.

Validation of new technologies

When we think of validation of new technologies we tend to exclusively think in terms of development and experimental validation *within* cryonics. A closer look at how new technologies are introduced in cryonics should lead to a more nuanced perspective. First of all, in some cases the scientific validation has already been done in mainstream science and clinical practice. In emergency medicine a routine procedure is to stabilize the patient for subsequent hospital admission and treatment. In cryonics we would like to stabilize the patient for long term care at low temperatures. In both cases, however, the aim is to prevent any further deterioration from the condition we find the patient in. If a new mechanical device can deliver more effective external chest compressions (and improve cerebral blood flow), then, everything else the same, this should translate into improved patient care in cryonics, too. The crucial part here is "everything else the same." One subtle problem that is often underestimated by medical professionals who are new to cryonics is that the conditions in which cryonics patients present themselves can be so distinctly different that a departure from standard emergency medical protocol is necessary. Thus, often mainstream technologies need to be translated into cryonics technologies and sometimes this even requires additional experimental research. In general, though, adaptation of new mainstream technologies can accelerate the progress in cryonics technologies.

Another area in which the need for conducting experimental research is often minimal is when the technological changes in question are primarily engineering challenges. A good example concerns efforts to increase the cooling rate during initial stabilization. It is well recognized that faster cooling rates during this phase confer a substantial benefit and are instrumental to keep the patient's brain viable. Any technology of internal or external cooling that can achieve this objective constitutes measurable progress. Or consider the development of computer-controlled perfusion that can optimize a perfusion protocol based on a number of chosen variables (pressure, cryoprotectant concentration, et cetera.)

When it comes to the core technologies in cryonics such as cryopreservation of the brain, however, there is no credible alternative to conducting experimental research in-house or contracting with other research labs. In an ideal world, prior to adaptation, new cryopreservation technologies would be independently verified in a number of labs using different animal models and the new technology would then be progressively implemented in cryonics with extensive data collection and analysis. It is indisputable that this is the gold standard in cryonics but at this point it cannot be claimed that all cryonics technologies have been validated with such rigor. The rationale for using technologies in cryonics has ranged from theoretical extrapolations from the scientific literature to the use of technologies that have been validated in peer reviewed publications.

Conducting experimental research to validate new technologies is a non-trivial affair for the typical cryonics organization. Funding that can be allocated to research often needs to compete with other priorities such as maintaining qualified staff and promotion. There is also the increased recognition that combining patient care and experimental research is not prudent, which necessitates either outsourcing research or establishing separate research facilities. New technologies often produce new research questions. For example, the adoption of vitrification solutions has greatly increased interest in investigating low toxicity cryoprotectants.

Implementation of new technologies

After identification and validation, the final step is implementation of a new technology. As discussed above, in cases where the technology is already in use in mainstream medicine, implementation often requires some kind of adaptation for use in cryonics. Another important element of implementation is creating documentation and the training of staff and contractors to use the new technologies. In some cases, the lack of required skills can complicate or delay implementation.

Validation and implementation are not always distinct phases. Often, the only way experimental evidence can be obtained about a new technology is to carefully introduce it in human cases, collect data, and revise the technology if necessary. The introduction of new technologies should always be followed by focused and repeated data collection to evaluate its efficacy and to determine whether the addition of this technology brings the cryonics organization closer to its ultimate goal of reversible cryopreservation.

The technological progress that has been made in cryonics is impressive, especially considering its science and limited scientific support. Unlike a field such as biogerontology, cryonics protocols can usually be tested in a relatively short time span and there is little dispute over what kind of problems need to be solved to achieve reversible cryopreservation. In the remainder of this article I will give a number of reasons (some of them intrinsic to cryonics) that have prevented more rapid technological progress in cryonics.

Obstacles to rapid technological progress in cryonics

Before I start with reviewing a number of causes it will be helpful to reiterate an earlier observation; the idea of technological progress in cryonics follows the recognition that reversible cryopreservation (or human suspended animation) is the ultimate goal of cryonics procedures and that we can evaluate cryonics cases with this framework in mind. This leads us to the first reason that can explain a slower pace of technological development.

No formal commitment to human suspended animation

Without a strong commitment to human suspended animation as a goal, a cryonics organization is at risk of becoming a freeze-and-repair operation that just goes through the routines without a framework to identify a route forward. While it can be argued that repair of the frozen brain is technically feasible and plausible, placing a critically ill patient in suspended animation leaves no doubt that the medico-legal status of a cryonics patient should be considered "alive." When human suspended animation is recognized as a formal goal, a cryonics organization can be judged by its efforts to close the gap between its current technologies and this goal.

No recognition of the concept of patient care

Closely related to establishing a formal commitment to human suspended animation is the recognition that the concept of patient care in cryonics is meaningful and allows for setting standards of care. For example, a cryonics organization can aim for keeping the brain viable by contemporary medical criteria during stabilization, prevent dehydration and freezing of the brain following cryoprotection and cooling, and eliminate fracturing during long term care by storing closer to the glass transition temperature. In each case, data need to be collected to determine to what degree these goals were achieved. Careful scrutiny of case data can lead to designing new research questions or pushing standards to an even more ambitious goal.

One of the most formidable challenges in the field of cryonics is that there is no direct feedback in a way that is obvious and recognizable for most people. There are no patients returning home after the procedure and the only way to determine whether a cryonics organization delivers care to the standards it is technically capable of is to collect data on cooling rates, take blood samples, perform viability assays on microliter brain tissue samples, inspect the brain for ice formation, and analyze CT scans after cooldown.

When a cryonics organization is deemed capable of producing reproducible outcomes in a typical cryonics case, the framework of suspended animation can then be used to identify new technological innovations that will further improve the level of patient care.

Competing priorities and financial constraints

Naturally, when there is no money available for research, or to fabricate or purchase the new technologies, a cryonics organization can remain in technological stasis. Technological innovation is important but can't be the only goal for a cryonics organization. A credible cryonics organization has the secure care of its existing patients as its most import goal. Even more time-consuming can be a high caseload, which can consume most of the time of technical and medical staff at the expense of technological innovation. As a general rule, most cryonics organizations also devote some resources to outreach and growth.

While it is correct that technological advances are usually passed on to members in the form of higher cryopreservation minimums, the fear of making cryonics too expensive for the average member has often delayed introducing new technologies. A good example is intermediate temperature storage. Replacing care at liquid nitrogen temperature for ITS systems will increase the cost of long term care (at least in its current incarnation). One way for a cryonics organization to ensure that research and technological development is not pushed below other priorities is to create a separate research fund and solicit targeted contributions. Cryonics organizations that enjoy generous financial support can also consider spinning off a separate research organization.

Lack of competent technical and scientific staff

For a cryonics organization it is important to recruit staff members who are scientifically literate and committed to technological innovation. This is not only important for staff members with technical responsibilities. When the whole staff of an organization shows strong support for technological progress it is possible to create a culture of scientific excellence. In contrast, if a cryonics organization lacks staff with solid scientific or clinical credentials, technological progress and good patient care will be compromised. This is also the case when staff members have formal scientific or medical credentials but show little initiative or are incompetent. Cryonics organizations are small and poor hiring decisions can have profound effects on the nature of the organization. Since it is usually easier to hire than to fire, such problems can be persistent and hard to reverse.

One risk in cryonics is that staff members who have excellent scientific credentials are recruited to work in other organizations and companies. As a consequence, the most technically savvy cryonicists are not employed in cryonics organizations. This potential development is another reason for a cryonics organization to spin off a separate research organization. In such a structure the finest minds in cryonics can devote their time to scientific and technological issues relevant to cryonics without being slowed down by other aspects of a cryonics organization.

A good example of a technology that is held back by the lack of enough medically qualified staff is field cryoprotection. In a sense, the idea of conducting cryoprotection on-site prior to shipping the patient to a facility first is as old as the idea of cryonics itself. Eliminating the prolonged ischemic times times associated with remote blood washout and patient shipment in favor of doing field cryoprotection near the location where the patient is pronounced legally dead would constitute a major improvement in patient care. Prolonged transport times on water ice are fundamentally incompatible with the aim of reversible cryopreservation. Unfortunately, only a handful of remote cryonics cases have been conducted as field cryoprotection cases. If field cryoprotection is done for all cases where this is technically preferable, substantial cost savings could be reaped as well. Making such a transition, however, would require that a cryonics organization always have access to case personnel or contractors who are competent at surgery and perfusion, and have good cryobiological knowledge.

High turnover of staff and leadership

When there is a high turnover of management and/or staff within a cryonics organization it is hard to make technological progress or conduct long-term research projects. New management and staff members may also have different perspectives about which technological developments to pursue and, as a consequence, R&D in progress is discarded or put on hold.

Closely associated with this is the loss of institutional knowledge. Having a broad and deep understanding of cryonics is important to identify and pursue new technological directions and evaluate the quality of care at an organization. Absent such (distributed) knowledge, a cryonics organization can remain in stasis or move in reverse. At the Alcor Life Extension Foundation there have been multiple cases in which the quality of care worsened relative to prior administrations or where routine technological procedures were (unconsciously) abandoned because of poor intuitional knowledge transfer. In a worst case scenario the cryonics organization does not know that it does not know and promotes itself as delivering excellent care and committed to technological innovation while mistakes and poor R&D are rampant.

Faulty commitment to cryonics

Faulty commitment may seem a strange problem for a cryonics organization to have. But it certainly was a problem in the early days, when some naïve businessmen perceived cryonics to be a get-rich-quick scheme, or otherwise had unrealistic expectations. The popularity of cryonics turned out to be not as high as projected, and funding to undertake and continue operations, including long term care, proved very limited. Baffled by the problems, most of these people left the field, sometimes being forced to abandon patients.

In more recent years cryonics organizations have faced a different kind of problem. Organizations such as Alcor and Suspended Animation can afford to pay market wages for most of their positions and wages above prevailing market values are not unheard of. As a consequence, seeking employment at a cryonics organization can be a rational course of action, regardless of any personal or professional interest in cryonics. In such a situation, a strong commitment to patient care and research is often lacking. Requiring staff to have cryonics arrangements in place is no longer a sufficient guarantee of dedication in these circumstances because obtaining cryonics arrangements can be considered just a small inconvenience for a well-paid job that lacks the usual professional scrutiny.

"The perfect is the enemy of the good"

One cause for a substantial delay between validation and implementation is to aim for a perfect technological solution before authorizing a technology to be used in cryonics. In reality this can mean that a technology that can already make a substantial contribution to patient care is withheld from the field. A prime example of such a technology, in my opinion, is liquid ventilation (or cyclic lung lavage). The feasibility and desirability of such a technology was established in the mid 1990s but at least 20 years has passed without formal deployment of this technology in cryonics despite various organizations having pursued its development. In fact, in this case a lot of the reasons for technological stasis in cryonics (such as high turnover of management and staff) seem to have colluded.

Another example may be intermediate temperature storage (ITS). If the recommended ITS temperature substantially reduces the amount of cracking but does not always eliminate it, a case can still be made for implementation this technology. This is particularly true if the brain is saved from fracturing events and the only remaining fractures can be healed through conventional surgery or organ replacement.

A related, but more subtle problem is not recognizing that a technology can be considered mature enough to make a contribution to cryonics but cannot be considered sufficiently developed for clinical use. A good example is organ vitrification. One might argue that the knowledge that sufficiently high concentrations of cryoprotectant can prevent ice formation existed for a long time in cryonics before it was introduced in the field. Since neither conventional cryopreservation nor vitrification could produce high viability readings, the only useful indicators for cryonics could have been inhibition of ice formation and histology. By these criteria even the vitrification solutions that did not produce good viability in slice work would have been a sensible replacement for the prevailing glycerol protocols.

Conclusion

Without formalizing reversible cryopreservation as a research and clinical goal, a cryonics organization is at risk of technological stasis and poorly positioned to identify, validate, and implement superior technologies that aim to close the gap between prevailing procedures and human suspended animation. Rapid technological progress in cryonics requires prudent hiring, a tech-savvy and scientifically literate staff, a stable culture committed to cryonics, a distinct R&D program, generous financial support, and the ability to prioritize technological needs based on research and observations made in casework.

Perhaps the most formidable obstacle to creating and sustaining such an infrastructure is the lack of obvious feedback in cryonics procedures. There is no revival or healing that can easily be understood by members and the general public. Thus there is only limited validation of, or motivation to insist on, good patient care and ongoing technological innovation. The vision that cryonics organizations should offer something better than store-and-repair has always had its advocates but its influence has remained limited and fragile.

If cryonics organizations would introduce liquid ventilation, field cryoprotection, and fracture free storage, there are three remaining technological challenges to achieving human suspended animation. These are (1) the design of a vitrification agent with no or negligible toxicity, (2) eliminating severe cryoprotectant-induced dehydration of the brain, and (3) optimum distribution of the cryoprotectant in whole body cases.

History of Cryonics—A Narrative Analysis of the Cryonics Magazin

Dirk Nemitz

Abstract

I present a first attempt to analyze available material on cryonics in order to identify relationships and background information through in-depth analysis of articles in a member magazine published by the Alcor Life Extension Foundation on a regular basis since 1981.

Many cryonicists are convinced that it is absolutely critical to find a way to unlock this information "What do cryonicists write for cryonicists to read?", This is an attempt to answer this question and make it available for new members, so that mistakes are not forgotten and, most importantly, not repeated. At the same time, an operation of the scale and inter-disciplinary nature of cryonics, always has room for improvements, in particular when it's run by such a small number of organization employees and volunteers.

As R. Michael Perry puts it very elegantly, speaking for cryonicists: *"We must keep doing what we are doing, keep trying to do it better, and look for ways and opportunities to interest others."*

Introduction

The field of cryonics up to today is a very small social movement. Despite their existence since the 1970ies, the two large cryonics providers in the United States claim to have about 1,000 members each (the organizations are Alcor Life Extension Foundation in Phoenix, Arizona and the Cryonics Institute in Detroit, Michigan). At a current world population of about 7.2 billion people, this amounts to about 0.00000028 per cent. While this small portion of the world population interested in cryonics has generated an extensive media coverage, including in tv shows, movies, documentaries, radio shows, newspaper and magazine articles, the coverage in the scientific literature is surprisingly low. There's no published research paper available which analyzes the motivation, background, behaviour or psychology of cryonicists.

There are good reasons to make an attempt to change this situation. Although this social group currently is rather small, growth rates in membership since about 2005 signal that this is about to change. At the same time, cryonicists hold unique views on the philosophy of life and related important concepts, in particular death. It would also be interesting to find out more about potential overlap with other existing and emerging philosophical and religious worldviews. How unique is the understanding of cryonicists related to death? How compatible are the ideas of cryonics with views of existing world religions? How much overlap exists with upcoming movements, such as transhumanism?

This conference contribution aims to make a first attempt to analyze available material. In order to identify relationships and background information, it asks the question "What do cryonicists write for cryonicists to read?", and attempts to answer this question through in-depth analysis of articles in a member magazine published by the Alcor Life Extension Foundation on a regular basis since 1981. This can give an important insight into how cryonicists see themselves, what they think other cryonicists would read, how the construct the self-narrative of the cryonics movement, and what events in cryonics history have influenced the movement substantively.

Material and methods (in experimental original papers)

The research is undertaken based on 49 issues of the Cryonics Magazine over a 10 year period, which includes all issues published between 2004 and 2013. The Cryonics Magazine is the member magazine of the Alcor Life Extension Foundation. All of the material used for this study is freely available online (Alcor Life Extension Foundation, 2015).

The 49 issues contain a total of 1,414 pages. The analysis was undertaken on a filtered subset of this data in order to cut out advertisements, table of content, cover pages and other non-substantive content. This left a total of 372 articles on 967 pages to be analyzed.

The analysis of the articles contains qualitative and quantitative elements. For simple quantitative statistics, the articles were categorized and the number of pages per category were analyzed. In addition, qualitative methods were used, based on the narrative analysis. Traditionally, narrative analysis is a research method in order to study how people create meaning and consistency in their lives by using narratives (Riessman, 1993). Here, the method was used in a very similar way, but slightly adapted in order to research how cryonicists create a meaning of cryonics as narratives in their writings for other cryonicists, and in particular in order to identify the most critical events in cryonics' history as perceived by cryonicists.

Results

The results give a clear indication of what writers of the Cryonics Magazine think that other cryonicists would be interested to read. About 72.07% of the pages are covered by six topics: organizational information, technology, book reviews, case reports, cryopreservation and history (Fig. 1).

Organizational information covers between 7 and 40% of each issue, and in total 24.30% of the pages over the 10 year period. This mainly includes updates from Alcor's CEO, other updates and information related to the organization, and Alcor member profiles. Articles related to cryopreservation cover 16.13% of the pages and the specific articles on cryopreservation case reports an additional 8.07%. It's interesting to mention that there are three years during which no case report was published, while in other years the case reports alone took up 18% of the magazine pages. Together, all of these articles related to cryopreservation make up almost another quarter of the publication volume.

Articles related to technology cover about 10.65% of the pages. Especially the section "Tech news" is a very interesting section and could form a study area of its own, in particular comparing today's reality with predictions made 10 years ago. For example, issue 2 in 2010 reported on a "Breakthrough in Ebola Treatment" through the use of small interfering RNAs. It's also noteworthy that the section was missing in the years 2011 and 2012. Of the remaining categories, articles related to history take up 6.51% and book reviews take up 6.41% of the magazine pages. Both categories have quite a large variance between years, with history covering between 0 and 13% and book reviews covering between 1 and 15%.

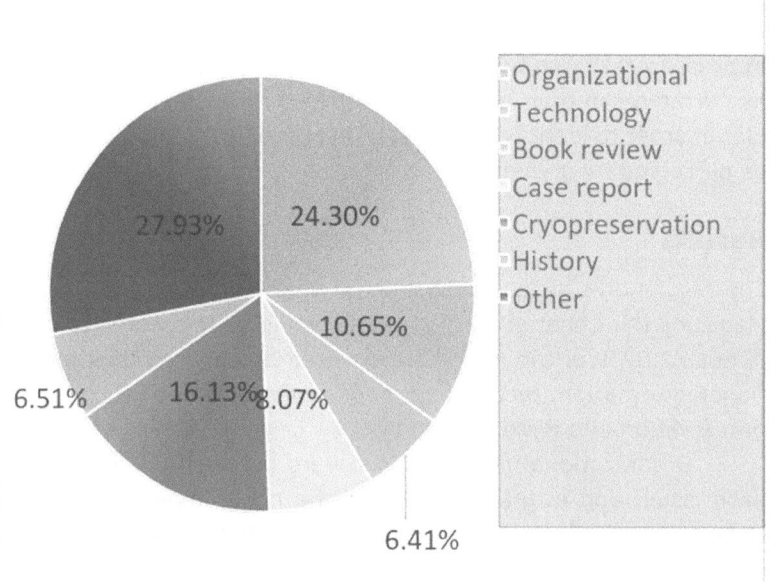

Fig. 1.
Percentage of page numbers by topic (967 pages = 100%)

The cut-off figure for dedicated categories was 5%, so the remaining topics included in the category "Other" cover less than 5% of page numbers each. The categories included are religion, events, ageing/immortality, legal issues, outreach, reanimation, singularity and transhumanism. In addition, this category also contains two poems.

On a more qualitative analysis, three important historic events came up recurrently which were outstanding and seemingly have impacted and influenced the cryonics movement substantively: the 1971 cryonics conference, the 1979 Chatsworth disaster, and the 1988 Dora Kent case. All three incidents will be shortly summarized here, while a deeper discussion of the related implications will be given in the discussion section.

The 1971 Cryonics Conference held in San Francisco was the subject of an extensive history article written by R. Michael Perry in issue 3 of the 2005 Cryonics Magazine volume. The Conference was organized by the Bay Area Cryonics Society (today renamed to American Cryonics Society). The article clearly indicates the importance of cryonics conferences, which had been organized on and off since 1963. The 1971 Cryonics Conference was the last conference of annual conferences held since 1968. It had more than 60 registered participants and fifteen papers were presented.

Interestingly, already at that time the talks and discussions covered a wide variety of interdisciplinary topics, including considerations related to resuscitation, cryonics research, technological improvements in cryopreservation, social and legal aspects of cryonics, life extension, financing cryonics, philosophy, nutrition and health.

R. Michael Perry concludes in his article that *"The year 1971 seems to have been one of those times when things looked better than they really were"*. The overall positive attitude towards the difference cryonics can make by providing a potential opportunity to avoid death was obvious during these early years of cryonics conference, but was soon challenged by an event which shocked the cryonics movement.

The most devastating event became known as the 1979 Chatsworth disaster. R. Michael Perry also provided an article on these *"Suspension failures: Lessons from the early Years"*, based on a 1992 article on the same topic and published in the first issue of the 2005 Cryonics Magazine volume.

The Chatsworth disaster resulted in nine cryonics patients thawed and lost and the owner of the for-profit business undertaking the cryopreservations was convicted of fraud and intentional infliction of emotional distress. At this point, the trust into cryonics operations was seriously challenged, because out of seventeen documented cryonics freezings undertaken until 1973, all but one had ended in failure. For at least another five cases, some of them privately maintained, the cryopreservation was later terminated. One of the main reasons for failure of the maintenance for early cryopreservation patients is limited and inadequate funding, often not provided as expected by relatives of the patient.

The interesting conclusion drawn by R. Michael Perry is that businesses and even family members uninterested in cryonics can't be entrusted with reliably running cryopreservation operations. In this field, those providing financial resources as well as those in charge of daily care should only be those who *"have strong personal interest in being cryopreserved themselves and have made arrangements."*

From these sad early events, two points are noteworthy. First, cryonicists are maintaining the memory of these events openly and transparently, despite the painful memories associated with them and the potential for negative assessment. It seems that this behaviour by itself is already increasing trust and ensuring that such mistakes may not happen again. Second, given the stability and structure of current cryonics organizations, and the lack of any cryosuspension failures in their operations since about 40 years, it can be concluded that *"cryonics seems to have entered a new era of strength, stability, and continued growth"*.

Another constant threat to cryonics operations resulted in the strengthening of the cryonics movement: the legal uncertainty that comes with such an experimental treatment, and associated lawsuits, in particular the 1988 Dora Kent case. Dora Kent was 83 old and gravely ill when she deanimated at Alcor, but no physician was present and the cryopreservation was started immediately. The absence of the physician started a coroner's investigation that turned increasingly hostile, resulting in accusations of homicide and practicing medicine without a licence.

What can clearly be derived from the readings also is that cryonicists firmly believe that their cause is just, and that many of them are dedicated. At the same time, there is a good reflection and understanding that their undertakings are not well understood by large parts of society, and may scare many people. As R. Michael Perry puts it: *"We must instead be careful and diligent at all times, and in particular try to minimize confrontations with an establishment that is powerful and could easily do us great harm while convinced it is doing good."*

Discussion

From the quantitative analysis results, a few conclusions can be drawn. The largest number of pages is taken up by organizational information, which shows that the organizations take it really seriously to address their members and update about latest organizational developments. It seems to be a priority for cryonics organizations to be transparent in how they're running day-to-day operations as well as long-term strategies. Given the number of member profiles and personal information released, it is also safe to assume that cryonicists like to read about other cryonicists, and getting to know more about others who share similar plans and intentions.

The curiosity and transparency also extends to articles dealing with cryopreservations, either in theory or in form of actual case reports. Cryonicists seem to get some kind of reassurance regarding the seriousness of the cryonics operator when reading about actual case reports, current short-comings and recent or future improvements to cryopreservation protocols.

The category "other" contained a wide variety of in total 19 topics with a coverage on less than 5% of pages. For example, this ranges from articles on legal consideration to letters to the editor and from the introduction of alternatives to cryonics to articles speculating on social change. For some of these categories, such a low coverage is rather surprising, e.g. there was only one article on religion (0.31% of pages), while the topic of religion is known to be controversially discussed among cryonicists at meetings, in mailing lists and online fora. Similar statements could be made in relation to transhumanism (0.21% of pages) and speculations related to potential future reanimation (2.90% of pages).

Surprisingly, also two poems were published, showing that the cryonicist movement is not limited to purely technical people and inputs. Namely, these were "Cryonics" in 2007 and "Isaac from the Outside" in 2011.

The qualitative analysis has been able to identify three events in cryonics history which significantly shaped the way the movement operates today. Cryonics conferences have been an important tool to bring together cryonicists since the very early days of the movement, at least back to 1963. It could be a very interesting analysis to try and find all of these events and compare their scope, extend and participation. It is also worthwhile to mention that some individuals have shown a remarkable consistency in their engagement for cryonics. For example, Professor Peter Gouras was one of the speakers of the 1971 Cryonics Conference in San Francisco mentioned above, but also spoke in 2014 at the second German Cryonics Symposium in Dresden, Germany—covering a time span of 43 years actively supporting cryonics, while also making a stunning career in the medical profession.

The other two events identified, the 1979 Chatsworth disaster and the 1988 Dora Kent case, were much more threatening at that time. The Chatsworth disaster is still recognized as a tragedy. It shattered belief and trust in for-profit organizations in the field of cryonics, and basically in any cryopreservation services provided by anyone who isn't a cryonicist with own cryopreservation arrangements firmly in place. The Dora Kent case was a legal threat to the operation of cryonics services in general. Through winning all associated court cases and being found not guilty in relation to all accusations, Alcor actually achieved a stronger legal status for organizations offering cryopreservation services. This outcome is widely recognized as being just, because there is no harm to any third person if someone takes the very personal decision to be cryopreserved upon death (Sethe 2013).

Finally, this study has been limited to a single magazine, and only ten volumes of this magazine, covering less than 20% of cryonics history. 967 pages analyzed quantitatively and qualitatively give a good first indication, but for deeper insights much more efforts could be undertaken. Other magazines, historic mailing lists and printed material could offer many more additional insights. Interesting questions could be differences between members of different organizations, stable and changing trends over time, or differences between cryonicists from different countries.

Conclusions

The first conclusion is that cryonicists are diligent readers and writers, and there is a whole wealth of information and experiences available which can help to understand the movement and its history better. It's noteworthy that the number of peer-reviewed journal papers related to cryonics is about a dozen in more than 50 years, while the media coverage seems to be far more extensive. The reasons for all of these could be part of further research.

The second conclusion is that the cryonics movement is particularly vulnerable, because in cryonics small formal or operational mistakes can have dire consequences, as seen in the Chatsworth disaster. Especially inadequate levels of funding and poor planning have led to failure of suspensions in the past, which seriously damaged the reputation of cryonics providers and let to the establishment of an environment where at least in the United States only non-profit organizations are successful in offering cryopreservation services.

Acknowledgements

My sincere gratitude goes to Klaus Sames, who has been mentoring me for all these years and is a continuous source of inspiration. I also thank Ben Best and Mike Perry for their generous support to this chapter, in particular by contributing material as well as sharing invaluable insights and background information. A large thank you also goes to Max More, whose 2014 symposium presentation in Germany and short but invaluable discussions helped to put the research at hand into perspective.

References

Alcor Life Extension Foundation. [Online] [Cited: October 5, 2015.] http://www.alcor.org/AboutAlcor/membershipstats.html

Alcor Life Extension Foundation. Cryonics Magazine. [Online] [Cited: October 5, 2015.] http://www.alcor.org/CryonicsMagazine/archive.html

Cryonics Institute. [Online] [Cited: October 5, 2015.] http://www.cryonics.org/ci-landing/

Riessman CK (1993) *Narrative Analysis.* SAGE Publications, Inc, Vol.30 Qualitative Research Methods. London (p. 80)

Sethe, S. 2013. Internal ethical issues in cryonics. In Sames K (ed.) *Applied Cryobiology—Human Biostasis.* Stuttgart: ibidem-Verlag 87–108

Molecular Repair at Physiological Conditions?

Klaus Mathwig

Abstract

The revival of patients in cryostasis requires substantial future developments in nanotechnology to repair tissue at a molecular level. Here, we assess state-of-the-art technologies for sensing of single molecules at physiological conditions, and point to challenges of molecular handling at ambient conditions in liquid.

Introduction

Reanimation of patients in cryonic suspension is not yet possible. Contemporary medical technology lacks the tools to reverse damage caused by the suspension process as well as by the original disease occurring before cryopreservation. When a patient is suspended under optimal conditions using modern vitrification solutions, no freezing damage occurs and tissue is preserved without any ice formation. Moreover, at the low temperatures of biostasis of −120°C or lower, tissue does not undergo any changes for an infinite amount of time. Nonetheless, substantial and sophisticated development in medical technology will be necessary in the future to enable resuscitation. This is due to a number of reasons: damage is caused by different origins, namely the suspension procedure and low temperature, the disease leading to cryosuspension in the first place, and aging. I.e., for a successful reanimation, it is not enough to just revive the patient, he needs to be cured too, and reversal of aging is highly desirable. This calls for a technology operating on very different length scales, it must be able to repair macroscopic damage such as cracks, cellular damage and molecular damage. Most diseases including aging are molecular diseases. Thus, to cure them it will be necessary to rearrange bonds of a vast number of individual molecules in a large number of cells.

Such an advanced repair technology does not exist yet, but the fundamental feasibility has been studied. It is widely anticipated that the reanimation of patients is cryonic suspension will be enabled by *molecular nanotechnology* (Drexler 1992). This technology would allow handling matter at an atomic scale by controlling molecules and assembling them at will. A likely implementation are nanorobots, a large number of these molecular machines would be able enter many cells simultaneously to repair molecules on a massively parallel scale. The potential benefits of nanotechnology for cryonics have been discussed already more then two decades ago (Merkle 1992); and several feasibility studies in nanomedicine and molecular nanotechnology have elaborated on possible implementations and procedures (Wowk 1988). These scenarios typically involve diamondoid nanorobots (Freitas 2010) operating on completely different principles than natural occurring molecular machines such as enzymes.

Repair during suspension?

Merkle and Freitas argue for a molecular repair that is taking place while the patient is still in suspension at lowest temperatures (Merkle and Freitas 2008). This is a conservative approach since any further damage and deterioration during and caused by the warming up can be completely avoided. Diamondoid nanorobots are anticipated to be fully functional also at lowest temperature as they rely on nanomechanical components such as gears and bearings which would work frictionless if fabricated with atomic precision; and they would be able to operate at any temperature.

A typical revival scenario would take place in several steps (Merkle and Freitas 2008): First, access is given to all cells via the veins, arteries and capillaries; nanorobots are used to remove material in the cardiovascular system. Thus, once such pathways are cleared, nanomachines can reach any point in the body and tissue within approximately 2 ug, and also communication and energy pathways to the outside resources are ensured. Nanodevices are then employed to assess damage and repair or stabilize larger microscopic fractures in tissue. Warming up starts slowly and fluid is reintroduced into the cardiovascular system.

Repair in cells starts and continues on a molecular level. This includes analysis and repair of denatured proteins and the supply of new proteins and other biomolecules as well as balancing the concentration of electrolytes, glucose, oxygen etc (Freitas Jr and Phoenix 2002). Whole cell compartments are being repaired while intracellular fluid starts to liquefy, and at higher temperatures stabilizing nanodevices are removed while organs regain their function.

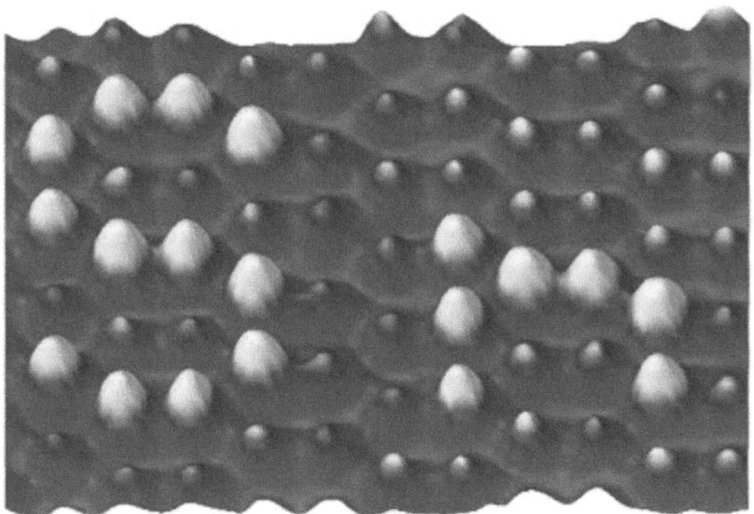

Fig. 1.
Tin atoms on a germanium surface are rearranged by 'near contract atomic force microscopy' at room temperature. The tin atoms form the chemical symbol "Sn". The image size is 8 nm x 8 nm. Reprinted by permission from Macmillan Publishers Ltd: Nature Nanotechnology (Sugimoto et al. 2005), copyright 2005.

Diamondoid nanorobots are still closer to the stage of feasibility studies and theoretical simulations than to practical implementation. Nonetheless, such a revival scenario might very well be possible in the future.

It is possible to position atoms at will with sub-nanometer precision since the 1980s and early 1990s (Eigler and Schweizer 1990). This groundbreaking research in nanotechnology has been achieved by atomic force microscopy (AFM). Here, an atomically sharp tip attached to a cantilever is moved over a surface so precisely that the composition of the surface structure is revealed by friction on the cantilever with atomic accuracy. Single atoms can be picked up by the cantilever tip and repositioned (see Fig. 1 and also Fig. 4a). The success of the AFM has inspired the conception and feasibility studies of molecular nanotechnology. The first AFM experiments have originally been performed at highly optimized and controlled experimental conditions of an ultra-high vacuum and lowest temperatures. Only much later, rearranging individual atoms became possible at room temperature (see Fig. 1).

While diamondoid nanorobots are far away from being a practically applied and fundamental research is still required, a diverse practical technological toolkit is being developed today to sense and also handle molecules at room temperature in liquid. Moreover, natural molecular machines, e.g., complex enzymes, operate in the same regime, and their functionality is being studied in greater and greater detail. Therefore, it could be possible that first successful reanimation of cryonics patients will involve molecular repair techniques operating at higher temperatures.

Here, we will point to the effects of Brownian motion and ionic screening which are of major importance for a possible molecular repair at *physiological conditions*, i.e., the conditions of a liquid at ambient temperature with a high concentration of electrolyte ions. These effects impede sensing and handling. We will summarize contemporary technological methods which nevertheless allow the detection and handling of single molecules.

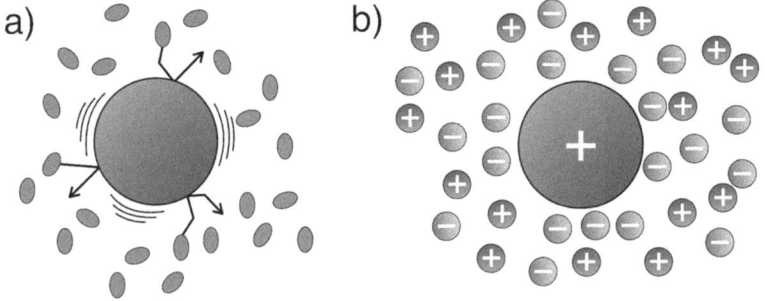

Fig. 2.
Effects in liquid at ambient temperature impede handling of single biomolecules. a) Brownian motion: water and other molecules constantly collide with target molecules. This leads to a random walk, i.e., a quick displacement of the molecule along a nanometer length scale and on a microsecond times scale. b) Electric screening: small counter ions in an electrolyte solution arrange around a (positively) charged target molecule to counterbalance and screen its charge on a nanometer length scale. This renders longer-range electrostatic handling impossible.

Brownian motion

Any molecule in liquid at room temperature undergoes *Brownian motion*, i.e., a completely undirected movement. On a microscopic level, temperature is defined as the movement of molecules. As water molecules move, they constantly collide with the target molecule (the analyte molecule targeted for handling or repair) constantly pushing it different directions—in a completely random fashion (see Fig. 2a). In a time t, each molecule undergoes a random walk, covering a distance l of

$$l = \sqrt{6Dt}.$$

Here D is the diffusion coefficient of the order 10^6 cm^2/s (typical for a small protein). Thus, a molecule will travel and be displaced about 50 000 nm in one second and still about 50 nm in a microsecond in a completely random unpredictable direction; even if free diffusion of a molecule is hindered by the crowded molecular environment of a cell and, thus, slowed down, the order of magnitude of the travelled distance still holds true. Therefore, Brownian motion is a considerable difficulty for handling molecules at room temperature, as the targets just escape very quickly.

Nonetheless, it is possible to overcome Brownian motion and trap molecules in solution using an Anti-Brownian Electrokinetic Trap (Cohen and Moerner 2005). In this microfluidic device the position of a single charged molecule in a microscope focus is monitored precisely by fluorescence microscopy. Electrodes are positioned around this focus and electrokinetic forces are generated (see Fig. 3a). As the molecule starts to diffuse out of the focus, the electrodes are actuated to establish an electric field and thereby to push the molecule back into the center of the device. Brownian motion is effectively overcome by feedback actuation, and this technique has been used to study more complex biomolecules (Wang et al. 2012).

Ionic Screening

The Anti-Brownian Electrokinetic Trap is an elegant way to confine a molecule or change its position; the molecule directly moves in an applied field by electrokinetic forces. However, generating these forces in a cell (possibly using miniaturized electrodes) would be very challenging. In any intercellular fluid, electrolyte ions are present in a very high concentration exceeding 100 mM. These high concentration leads to a strong *ionic screening*: For a negatively charged target molecule (such as, e.g., a protein), positively charged electrolyte ions will be attracted, surround it, and counterbalance the negative charge so that – viewed from some distance—the net charge is zero (see Figure 2b). The distance (Debye length) on which electrostatic interactions take place severely limits the length scale on which an electric field can be directly used for manipulation. This length κ^{-1} scales as $\kappa^{-1} \propto \sqrt{1/c}$, where c is the electrolyte concentration. At a typical electrolyte concentration of on the order of 100 mM, κ^{-1} is confined to 1 nm, making direct electrostatic handling impossible. Molecules become 'invisible' for sensing with electric fields.

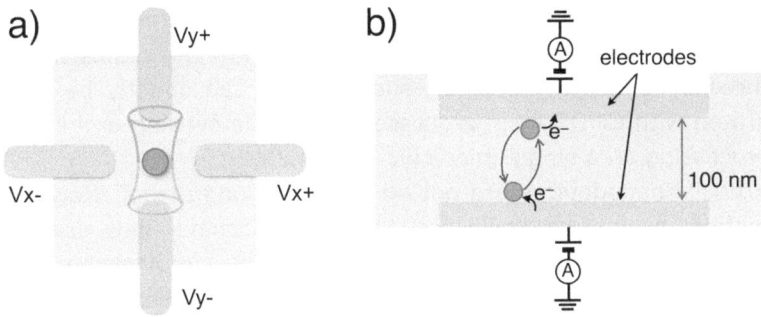

Fig. 3.
a) Schematic of an Anti-Brownian Electrokinetic Trap for handling single molecules in solution. A feedback loop is used in which the position of a molecule is determined optically and it is pushed towards the center between electrodes using electrokinetic forces. b) Schematic cross section of an electrochemical nanogap device. Utilizing Brownian motion, a single electrochemically active molecule is sensed as it diffuses back and forth between two electrodes where it is repeatedly oxidized and reduced, thus generating an electrical current across the nanogap.

Nonetheless, nanostructured devices can be used for the all-electrical sensing of single molecules. One way to make this possible is to use redox-active molecules, which directly exchange electrons when the collides with an appropriately biased electrodse by oxidation or reduction (Mathwig et al. 2014). While the charge of a single electron cannot be sensed at room temperature, Brownian motion can be utilized (Krause et al. 2014) to greatly amplify the charge which is generated by a single molecule. In electrochemical nanogap devices, an analyte molecule undergoes a random walk between to electrodes biased with an oxidation and reduction potential respectively (see Fig. 3b). The two electrodes form the floor and ceiling of a nanochannel; they are separated by a 100nm distance, which a molecule crosses within microseconds by diffusion. Therefore, each molecule can shuttle several thousands electrons per second across the nanochannel and generate a measurable current.

Nanogap sensors are among the devices manufactured by photolithographical microfabrication (Rassaei et al. 2011). Exactly the same tools are used for their fabrication which are used for microprocessors. All nanodevices manufactured by cleanroom microfabrication benefit from the advantages of an exponentially improved cost-effectiveness, miniaturization and massive parallelization, which is going to be extended into the nanoscale in three dimensions. Moreover, such sensors can directly be integrated with CMOS microprocessors to enable sensing and signal processing on a single microchip (Singh 2015).

Artificial nanodevices are not unique in making use of Brownian motion. Also in naturally occurring nanomachines it is used or rectified to generate directed motion. For example, Aktin and Myosin proteins rely on Brownian motion in the contraction of muscles (Kitamura et al. 1999).

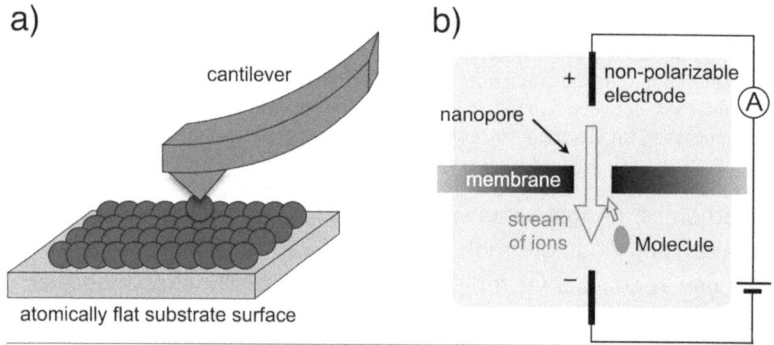

Fig. 4.
Schematic of single molecule handling and sensing. a) In an atomic force microscope, an atomically sharp tip attached to cantilever is used to pick up a single atom from a surface and redeposit it. b) Nanopore sensing. A change in current is recorded as a single analyte molecule traverses a nanopore and blocks the ion current.

State-of-the-art methods for single molecule sensing

In addition to above-mentioned techniques, a larger nanotechnological single-molecule toolkit exist.

The most widely used technology to sense molecule in biology by far is optical imaging by light microscopy, especially by super-resolution techniques. These methods encompass a wider range of techniques and functional principles which have been developed and applied exponentially in the past years (Huang et al. 2009; Galbraith and Galbraith 2011). They all have in common that they allow optical imaging with a very high spatial resolution beyond the diffraction limit and enable routine imaging in cell tissue at the level of single (fluorescently labeled) molecules. Nonetheless, is not clear how these techniques might facilitate molecular repair in the future as they are used exclusively for imaging but not for handling and as it is unlikely that a microscope can be shrunk to a nanoscale size. The same limitation apply to Anti-Brownian Electrokinetic Traps: while the microfluidic trap itself is a microscale device, it relies on the integration with a microscope which would be extremely challenging to miniaturize.

After the invention of atomic force microscopy in the 1980s, this technique has only be more recently been employed at room temperature (Sugimoto et al. 2005) or in aqueous solution (Roiter and Minko 2005). In this experiment in liquid, it was possible to observe the conformation of single electrolyte chains. However, these molecules were still adsorbed on a solid surface, i.e., they did not undergo a three dimensional Brownian walk. Nonetheless, while a complete microscope is still large, the actual positioning cantilever is only micron-sized, and atomic force microscope can be further miniaturized and parallelized.

In analytical (bio)chemistry, nanopore technologies have been established in the last decades as a sophisticated tool for single molecule detection (Howorka and Siwy 2009). This method is based on sensing of a molecule as is traverses a nanopore which separates two liquid-filled reservoirs. The nanopore is a nanometer sized orifice—either microfabricated or a biological nanopore (alpha hemolysin) in, e.g., a lipid bilayer. A voltage difference is applied to both reservoirs using non-polarizable electrodes such as Ag/AgCl electrodes. These electrodes consume electrolyte ions; they drive and sense an ionic current; its magnitude is determined by the bottleneck electrical resistance of the small nanopore. As target analyte molecules are driven through the pore by diffusion or electrophoresis, they exclude the conducting ions. The ionic current is blocked and fewer ions can contribute to the detected signal. Nanopore sensors are investigated as tools for DNA sequencing. Here, the genetic code would be read out as a single DNA molecule is pulled though the pore and the detected signal changes as a function of the nucleotide which is traversing the pore (Branton et al. 2008).

While nanopore device are powerful tools for sensing and the translocation of the molecule can be influenced by the potential applied across the pore, there is typically no direct control. The entrance into the pore as well as the speed of translocation is still a stochastic process. Nonetheless, when combined with optical methods, it is possible to mechanically pull a DNA with precise control. This can be accomplished by attaching one end of the DNA to a dielectric microscopic bead, which is then trapped in the intense light field of a laser beam. By moving the beam focus the DNA is then pulled through the nanopore with a high degree of control (van Dorp et al. 2009).

Conclusion

When the molecular repair of tissue of patients in cryonics suspension is conducted at room temperature, the effects of Brownian motion due to temperature and ionic screening hinder sensing of handling of single biomolecules. Nonetheless, new techniques are emerging which circumvent these difficulties and enable handling of single molecules at room temperature in electrolyte solution in a number of different ways. It is probable that the reanimation of cryonics patients will be possible with further future developments in nanotechnology. However, it is not decided yet which of these techniques can be miniaturized sufficiently and be employed in a highly parallel way. It is also not clear at which temperature repair will be conducted. A combined approach seems likely in which different procedures at different temperatures during warm up and reanimation might prove successful.

References

Branton D, Deamer DW, Marziali A, et al. (2008) The potential and challenges of nanopore sequencing. Nat Biotechnol 26: 1146–53

Cohen AE, Moerner WE (2005) Method for trapping and manipulating nanoscale objects in solution. Appl Phys Lett 86: 093109

Drexler KE (1992) Nanosystems: Molecular Machinery, Manufacturing, and Computation. John Wiley & Sons, Inc., New York, NY, USA

Eigler DM, Schweizer EK (1990) Positioning single atoms with a scanning tunnelling microscope. Nature 344: 524–6

Freitas Jr RA, Phoenix CJ (2002) Vasculoid: A personal nanomedical appliance to replace human blood. J Evol Technol 11: 1–139

Freitas RA (2010) Comprehensive Nanorobotic Control of Human Morbidity and Aging. In: Fahy GM, West DMD, Coles LS, Harris SB (eds) The Future of Aging. Springer Netherlands 685–805

Galbraith CG, Galbraith JA (2011) Super-resolution microscopy at a glance. J Cell Sci 124: 1607–11

Howorka S, Siwy Z (2009) Nanopore analytics: sensing of single molecules. Chem Soc Rev 38: 2360–84

Huang B, Bates M, Zhuang X (2009) Super-Resolution Fluorescence Microscopy. Annu Rev Biochem 78: 993–1016

Kitamura K, Tokunaga M, Iwane AH, Yanagida T (1999) A single myosin head moves along an actin filament with regular steps of 5.3 nanometres. Nature 397: 129–34

Krause KJ, Mathwig K, Wolfrum B, Lemay SG (2014) Brownian motion in electrochemical nanodevices. Eur Phys J Spec Top 223: 3165–78

Mathwig K, Aartsma TJ, Canters GW, Lemay SG (2014) Nanoscale Methods for Single-Molecule Electrochemistry. Annu Rev Anal Chem 7: 383–404

Merkle RC (1992) The technical feasibility of cryonics. Med Hypotheses 39: 6–16.

Merkle RC, Freitas RA (2008) A Cryopreservation Revival Scenario Using Molecular Nanotechnology. Alcor Cryonics 4th Quarter 28 online http://www.alcor.org/Library/html/MNTscenario.html

Rassaei L, Singh PS, Lemay SG (2011) Lithography-based nanoelectron-chemistry. Anal Chem 83: 3974–80

Roiter Y, Minko S (2005) AFM Single Molecule Experiments at the Solid−Liquid Interface: In Situ Conformation of Adsorbed Flexible Polyelectrolyte Chains. J Am Chem Soc 127: 15688–89

Singh P (2015) From Sensors to Systems: CMOS-Integrated Electrochemical Biosensors. IEEE Access PP:1–1

Sugimoto Y, Abe M, Hirayama S, et al. (2005) Atom inlays performed at room temperature using atomic force microscopy. Nat Mater 4: 156–9

Van Dorp S, Keyser UF, Dekker NH, et al. (2009) Origin of the electrophoretic force on DNA in solid-state nanopores. Nat Phys 5: 347–51

Wang Q, Goldsmith RH, Jiang Y, et al (2012) Probing Single Biomolecules in Solution Using the Anti-Brownian Electrokinetic (ABEL) Trap. Acc Chem Res 45: 1955–64

Wowk B (1988) Cell repair technology. Alcor Cryonics Vol 21–30 online http://alcor.org/Library/html/cellrepairmachines.html

Anti-Ageing and Pro-Longevity: What can We Learn from a Small Worm? A Methodical Overview

Nadine Saul

Abstract

The pursuit of eternal youth and immortality is an old dream. As far back as 2000 years ago, Alexander the Great was searching for the mystic fountain of youth. Even though scientific efforts are more realistic nowadays, researchers can demonstrate initial success with model organisms. Especially one animal enjoys great popularity among biogerontologists: the nematode *Caenorhabditis elegans* (*C. elegans*). This about 1 mm sized worm has excellent properties for research, including simple cultivation, a short mean lifespan of about 15 days and a relatively undemanding existence. *C. elegans* also has an astonishing genetic analogy to humans, thus it has been utilized to study the ageing process, as well as complex ageing-related diseases like Alzheimer's, cancer and diabetes.

At least four methods to retard the ageing process and to significantly prolong the lifespan of *C. elegans* are known today:
- Hormesis: Longevity by a mild chemical or physical stress
- Calorie Restriction (CR): Longevity by drastically reducing the amount of food
- Targeted molecular modulation: Longevity by specific gene deactivation
- Deep freezing: Longevity by freezing at −80°C or in liquid nitrogen

Hormesis and CR usually only lead to moderate lifespan extension and the effectiveness in humans is still under discussion. Moreover, some studies indicate that side effects are very likely, thus these two methods are probably not suitable for the application in humans.

The mutation of specific genes can lead to five-fold extension of lifespan in *C. elegans*, which equates 400–500 years in humans. And when using a cryoprotectant, about half of the nematodes survive deep freezing even for several years or decades. Although these two procedures are very promising, several questions regarding ethical, technical and biomedical problems need to be addressed before usage in humans is possible. In the end one will see whether humans are "only big worms" with similar success.

1. Why do we use the nematode *C. Elegans* ?

The nematode *C. elegans* is only 1 mm in length and we need a microscope to observe it (Fig. 1A). So why are so many researchers interested in such a small worm? There are numerous reasons as follows:

- Easy handling & cultivation: This nematode can live on small, space-saving agar plates which contain all essential nutrients. Furthermore, *C. elegans* uses *E. coli* bacteria as food source which are easy to cultivate in any lab.
- Hermaphroditic reproduction & high number of progeny: The offspring of each worm are clones which offers genetic uniformity and is an advantage in lifespan studies. Moreover, one animal produces about 300 eggs in a few days.
- Short lifespan & generation time: Biogerontologists like to use *C. elegans* due to its short lifespan which is about 15 days (median lifespan) or 30 days (maximum lifespan), respectively, at 20 °C (Fig. 1B). Thus, one lifespan assay is quite fast compared to mammal assays.
- Shows typical signs of ageing: *C. elegans* and humans share the most important signs of ageing:
    - *Less and slower movement*
    - *Decreased muscle function*
    - *Less stress resistance*
    - *Increased danger of infection*
    - *Decreased protein homeostasis*
    - *Cognitive decline*

- **Diverse mutant strains:** Numerous strains with one (or more) missing or dysfunctional gene(s) are available. When a life-extending substance is not working in a mutant strain, it is a clue for this gene being involved in the life-extending mechanism of this substance (Fig. 2).
- **Complete sequenced genome:** *C. elegans* was the first animal whose genome was completely sequenced. Thus, whole genome microarray experiments are possible. Furthermore, more than 80% of the *C. elegans* proteom have human homologues (Lai et al. 2000); therefore, the obtained results will also give important clues for human issues.

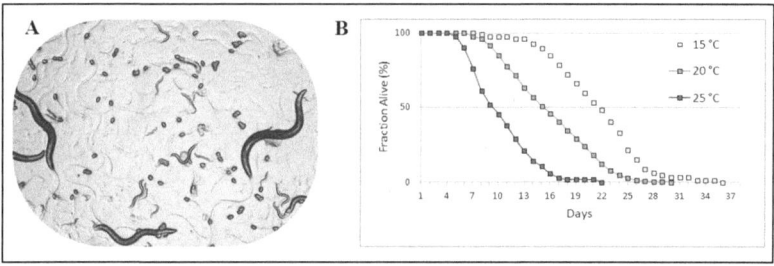

Fig. 1.
C. elegans and its typical lifespan curves
Shown is the microscopic view onto a plate with *C. elegans* in egg, larval, and adult stages (A) and lifespan curves of *C. elegans* at three different temperatures (B).

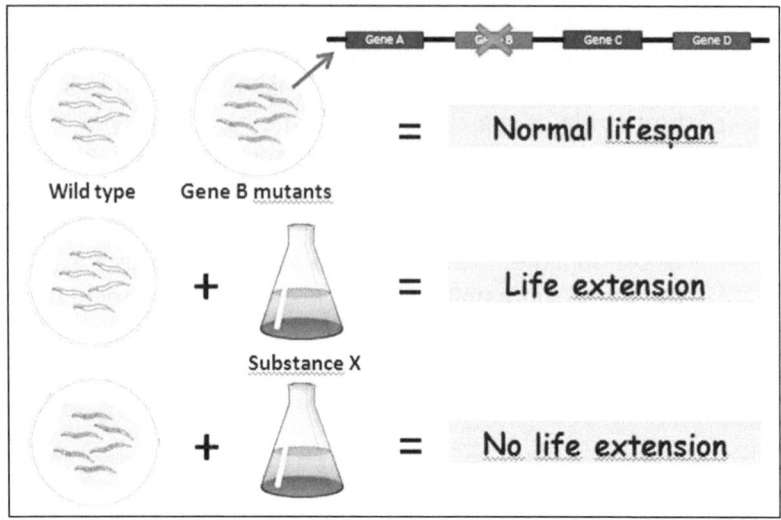

Fig. 2.
Mutant strains in lifespan studies
The life-prolonging feature of a test-substance is missing in a hypothetical gene B mutant strain. Thus, gene B could play an important role in the life extension mechanism of Substance X.

2. Why do we use *C. elegans* microarrays?

All genes (more than 22,000) of *C. elegans* are represented in several specific copies on a small microarray chip (Fig. 3A). Usually, sequences of about 25 bases per gene are spotted on a 1,28 x 1,28 cm glass slide.

In a typical microarray experiment, there are two groups of animals which should be compared: One group which was treated with, for example, a life-extending substance and the control group (Fig. 3B, 3C). Some substances can change the gene expression pattern by enhancing or inhibiting the transcription of single genes, resulting in a decreased or increased amount of specific RNAs and probably also in a changed amount of the corresponding protein. Therefore, the RNA of both groups will be collected, labeled with a specific color and hybridized with the probes on the chip. Via complementary bounds, the RNA of each gene will stick to its specific sequence on the chip. Finally, the labeling will be scanned and the RNA of each gene can be quantified. The resulting gene expression pattern can give valuable information about the mode of action of a substance and its longevity capacity.

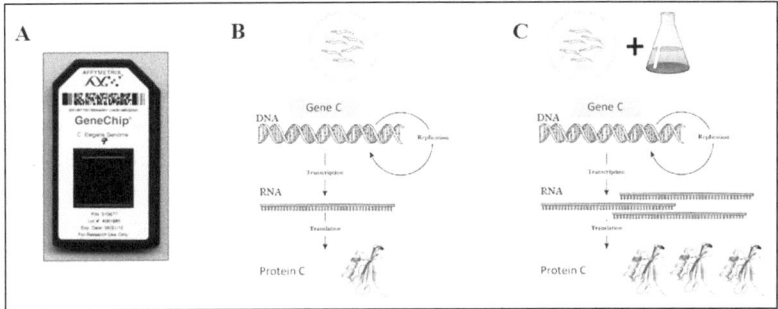

Fig. 3.
A *C. elegans* microarray chip and the experimental background

Shown is an example for a microarray chip (A) which represents more than 22,000 transcripts. Moreover, a typical experimental setup is shown with a control group (B) and a treated group (C). Here, the amount of RNA of a hypothetical gene C is increased by the test-substance.

3. Life extension in *C. elegans* and humans

The main challenges in biogerontology are to prolong the lifespan and to retard the ageing and healthspan-decline, which are two different goals, since they are not mandatory linked to each other (Avanesian et al. 2010; Garcia-Valles et al. 2013; Wanagat et al. 2010). Finally, the data should be transferred from the model organisms to humans. At least four proven mechanisms for life extension do exist for *C. elegans* so far: Hormesis, calorie restriction, targeted molecular modulation, and deep freezing.

3.1 Hormesis

Hormesis is a mild stress like a toxic substance in low concentrations with beneficial effects on an organism, which could be a longer life, an enhanced stress resistance or an improved reproduction. Hormesis is assumed to be independent of the endpoint, organism and stressor and is characterized by an inverted Ushape dose-response curve (Fig. 4) (Calabrese 2010; Calabrese and Baldwin 2001; Douglas 2008). In *C. elegans* hormesis was observed with radiation, inorganic & organic substances and heat stress (Cypser and Johnson 2002; Cypser et al. 2006). Mostly, an adaptive stress response is discussed as possible background mechanism for hormesis, which is based on the training of the stress defense system by a small amount of stressor. However, recent studies raise doubts concerning this theory (Steinberg et al. 2013).

Even with natural polyphenols, like ellagic acid or tannic acid, hormetic lifespan extension was observed. However, the life extending concentrations of these substances also led to detrimental side effects, like a delay in the initial reproduction, reduced growth or decreased oxidative stress resistance (Saul et al. 2013). Thus, this is further evidence that a longer lifespan is not necessarily connected to an improved healthspan.

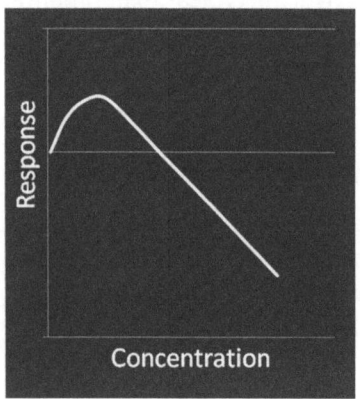

Fig. 4.
A typical hormesis curve

3.1.1 Hormesis in humans?

Is hormesis possible in humans? There are several examples of hormesis in observational human studies (also nicely reviewed in Mattson 2008a) like:
- Slight radiation stress leads to cancer protection (Kant et al. 2003)

- Slight ischemia stress leads to protection against heart attack (Yellon and Downey 2003)
- Exercise stress improves disease resistance (Radak 2008)
- Thinking stress improves brain function (Mattson et al. 2002)

Thus, in principle the hormetic effect is also working in humans. However, several disadvantages need to be mentioned:

- Dangerous effects with antibiotics and chemotherapeutics were reported, which can have hormetic effects regarding the pathogen or cancer cell, respectively, when they are applied too short or in low concentrations (Calabrese 2005; Davies et al. 2006).
- There is only a narrow concentration range of the hormetic effect. Thus, if too much of the stressor is used, toxic effects will appear.
- Only a slight life extension is possible via hormesis.
- Side effects, as described for the polyphenolic study, are possible.
- Longevity via hormesis is not proven in humans, yet.

3.2 Calorie Restriction

Calorie Restriction (CR) leads to longevity and ageing deceleration by reducing the amount of food to about 30–60% of the *ad libitum* amount. One theory about the background mechanism is, that CR is a kind of mild stress and acts in a hormetic way (Mattson 2008b). CR increases the life expectancy in almost all species. For instance, feeding *C. elegans* only up to the 6th day of adulthood (which assures correct development and reproduction), and leaving the nematodes completely without food thereafter, nearly leads to a doubled median lifespan (Fig. 5). Furthermore, experiments with rhesus monkeys, which got about 30% less food than usual, had a better brain function and featured less age related diseases like cancer, heart diseases, or diabetes. However, it is still controversial if this diet also leads to a longer lifespan in the monkeys (Colman et al. 2014).

3.2.1 Calorie Restriction in humans?

It is assumed that a diet in terms of CR is also working in humans, since several protective effects were reported, like protection against obesity, type 2 diabetes, inflammations, hypertension, cardiovascular diseases, and cancer (Rizza et al. 2014). However, side effects reported in model organisms as growth, development, or reproduction dysfunction are also under discussion for humans. Moreover, the capacity of CR to prolong the lifespan in humans is not proven, yet. Finally, humans might suffer from an impaired quality of life during such a strict diet.

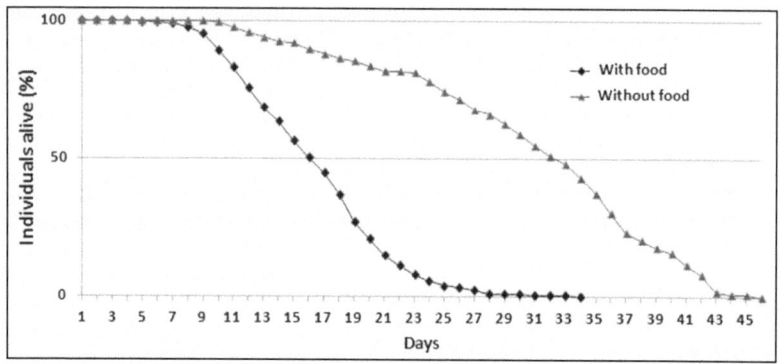

Fig. 5.
Longer life without food in *C. elegans*
C. elegans was fed the whole life (blue) or only up to the 6th day of adulthood (red). The median lifespan is indicated by the line at 50%.

3.3 Targeted molecular modulation

Most molecular modulations in *C. elegans* leading to life extension are part of one of the following pathways (Torgovnick et al. 2013):
- Insulin/IGF-1 signalling
- Mitochondrial signalling
- Germline signalling
- Neuronal signalling
- Food intake

For example, the mutation of two genes (*daf-2* and *rsks-1*) leads to a 5-fold life-extension in *C. elegans* (Chen et al. 2013). Mathematically, this corresponds to a lifespan in humans up to almost 500 years! However, even if molecular modulations result in very robust and reproducible effects in model organisms, the usage of this method is questionable due to the following reasons:

- Healthspan impairment is possible since numerous mutant strains in *C. elegans* with prolonged life showed severe healthspan decline in parallel, like growth, reproduction or movability restrictions (Bansal et al. 2015)
- Targeted mutations in humans are not possible yet due to huge ethical concerns
- It is not proven that specific mutations lead to longevity in humans
- Usually, targeted DNA mutations will be only active in the offspring, not in the individual itself. However, new methods like RNA interference (RNAi), which results in specific gene silencing by removing its corresponding RNA, could work directly in the individuals (Battistella and Marsden 2015). This method is currently surveyed.

3.4 Deep freezing

C. elegans is one of the most developed animals which can be easily deep frozen up to several years or even decades. Using the right freezing procedure, survival rates up to 85% are possible: Worms should be rapidly cooled to −3 °C with an ice seeding treatment (to avoid spontaneous nucleation) and 5% DMSO as a cryoprotectant. Thereafter, a cooling rate of 0.2 °C per minute should be conducted up to −100 °C and finally the worms should be stored in liquid nitrogen. Slow warming even after years will then result in a survival rate of about 85% for larvae and adult animals.

Hayashi et al. (2013) compared different ice seeding treatments, namely steel needles which are pre-cooled with liquid nitrogen, the addition of *Pseudomonas syringae* and a control without any ice seeding treatment. Both ice seeding protocols resulted in similar survival rates and showed an advantage compared to the untreated control. Moreover, they also compared different cooling rates in the presence of *P. syringae*. After initial freezing to −3 °C, they cooled the nematodes to −100 °C with 3.2, 0.5 and 0.2 °C/min, respectively. It was shown that slowing down the freezing will enhance the survival rate.

Now we need to face the challenge of finding the right freezing procedure for bigger animals and finally for humans.

References

Avanesian A, Khodayari B, Felgner JS, Jafari M (2010) Lamotrigine extends lifespan but compromises health span in *Drosophila melanogaster*. Biogerontology 11: 45–52

Bansal A, Zhu LJ, Yen K, Tissenbaum HA (2015) Uncoupling lifespan and healthspan in *Caenorhabditis elegans* longevity mutants. Proc Natl Acad Sci U S A 112: E277–86

Battistella M, Marsden PA (2015) Advances, nuances, and potential pitfalls when exploiting the therapeutic potential of RNA interference. Clin Pharmacol Ther 97: 79–87

Calabrese EJ (2005) Cancer biology and hormesis: human tumor cell lines commonly display hormetic (biphasic) dose responses. Crit Rev Toxicol 35: 463–82

Calabrese EJ (2010) Hormesis is central to toxicology, pharmacology and risk assessment. Hum Exp Toxicol 29: 249–61

Calabrese EJ, Baldwin LA (2001) U-shaped dose–responses in biology, toxicology, and public health. Annu Rev Public Health 22: 15–33

Chen D, Li PW, Goldstein BA et al. (2013) Germline signaling mediates the synergistically prolonged longevity produced by double mutations in *daf-2* and *rsks-1* in *C. elegans*. Cell Rep 5: 1600–10

Colman RJ, Beasley TM, Kemnitz JW et al. (2014) Caloric restriction reduces age-related and all-cause mortality in rhesus monkeys. Nat Commun 5: 3557

Cypser JR, Johnson TE (2002) Multiple stressors in *Caenorhabditis elegans* induce stress hormesis and extended longevity. J Gerontol A Biol Sci Med Sci 57: B109–B14

Cypser JR, Tedesco P, Johnson TE (2006) Hormesis and aging in *Caenorhabditis elegans*. Exp Gerontol 41: 935–9

Davies J, Spiegelman GB, Yim G (2006) The world of subinhibitory antibiotic concentrations. Curr Opin Microbiol 9: 445–53

Douglas H (2008) Science, hormesis and regulation. Hum Exp Toxicol 27: 603–7

Garcia-Valles R, Gomez-Cabrera MC, Rodriguez-Mañas L et al. (2013) Life-long spontaneous exercise does not prolong lifespan but improves health span in mice. Longev Healthspan 2: 14

Hayashi M, Amino H, Kita K, Murase N (2013) Cryopreservation of nematode *Caenorhabditis elegans* in the adult stage. Cryo Letters 34: 388–95

Kant K, Chauhan RP, Sharma GS, Chakarvarti SK (2003) Hormesis in humans exposed to low-level ionising radiation. Int J Low Radiat 1: 76–87

Lai CH, Chou CY, Ch'ang LY et al. (2000) Identification of novel human genes evolutionarily conserved in *Caenorhabditis elegans* by comparative proteomics. Genome Res 10:703–13

Mattson MP (2008a) Hormesis defined. Ageing Res Rev 7: 1–7

Mattson MP (2008b) Dietary factors, hormesis and health. Ageing Res Rev 7: 43–8

Mattson MP, Duan W, Chan SL et al. (2002) Neuroprotective and neurorestorative signal transduction mechanisms in brain aging: modification by genes, diet and behaviour. Neurobiol Aging 23: 695–705

Radak Z (2008) Exercise, oxidative stress and hormesis. Ageing Res Rev 7: 34–42

Rizza W, Veronese N, Fontana L (2014) What are the roles of calorie restriction and diet quality in promoting healthy longevity? Ageing Res Rev 13:38–45

Saul N, Pietsch K, Stürzenbaum SR et al. (2013) Hormesis with tannins: free of charge or cost-intensive? Chemosphere 93: 1005–8

Steinberg CE, Pietsch K, Saul N et al. (2013) Transcript Expression Patterns Illuminate the Mechanistic Background of Hormesis in *Caenorhabditis elegans* Maupas. Dose Response 11: 558–76

Torgovnick A, Schiavi A, Maglioni S, Ventura N (2013) Healthy aging: what can we learn from *Caenorhabditis elegans*? Z Gerontol Geriatr 46: 623–8

Yellon DM, Downey JM (2003) Preconditioning the myocardium: from cellular physiology to clinical cardiology. Physiol Rev 83: 1113–51

Wanagat J, Dai DF, Rabinovitch P (2010) Mitochondrial oxidative stress and mammalian healthspan. Mech Ageing Dev 131: 527–35

Definitions of Death

Klaus H Sames

Abstract

Death is the most horrifying human experience. Biologically there exists general agreement that brain damage occurs gradually during total ischemia following organ failure. For up to 8 hours or more following heart arrest many neurons are viable, can be cultured, and can regain electrical activity. Lack of energy explains the fact that apoptosis remains incomplete and macrophage action is inhibited. Therefore, brain structures are preserved for a long time and may, furthermore, contain many active stem cells able to differentiate into neurons. In the absence of extended cryopreservation damage, there exists a chance to repair the brain following longer ischemia and vitrification.

Introduction

Death is no subject but a negation meaning absence of life. Dying and the following unlimited absence of life are an imagination, which cannot be faced without heavy emotional reactions. In medicine some different definitions of death describe the somatic transition to non-being.

The psychosocial aspect

Facing death

A German psychiatrist has specified the feelings arising in the face of death to be—fear, shock, helplessness, grief, resignation, and despair—the most severe and most negative feelings a human being can experience (Fuchs 2001).

Fear may be provoked by thoughts of concrete images, such as pain during dying or even during death as well as after death. Other emotions may be raised by the unimaginable, such as non-existence, no return forever, inevitability, or helplessness.

To understand coping with death we are to start early in life. If the beloved pet of an infant is dead, he/she runs to his/her parents begging, pleading, and urging them to revive the animal. In the course of time, he/she will learn that death is irreversible. Again some years later the infant faces his/her own death. It seems far away; it happens to adults, to very old people ...... sometimes. Full realization of one's own mortality follows around the beginning of puberty. The teenager does what a teenager always does. If he/she feels molested, he/she rebels.

But at the end stands resignation and even acceptance of death while trying to look for its friendly side. This may take place around the age of 19 (see Bürgin 1981). Thereby, a conception of arrangement with death may be formed that "crystallizes," resulting in a loss of flexibility and avoiding further confrontation with death. Moreover, such conceptions may become fixed by tradition, for example, the religion of the family.

How can a person arrange oneself with death? One way is religion. A religious person may believe in the concept of paradise. Religious people, who do not believe in metaphysics, may tell you that there must be a power that has brought them into life, and this power, may be named God, will not destroy his own work.

Another refuge is society. Societies seem to be immortal, and finding one's place in the culture of a society is a way to participate in its immortality.

It is hard to understand people, who tell us that their own extinction, an extinction forever, does not matter to them. One can explain this only in terms of suppression of all thoughts and feelings involving death. However, psychiatrists warn of depression of emotions, which may produce pathopsychology or even psychiatric diseases. Moreover, the proposed antidote to face death a realistic way will not function in this context (Fuchs 2001; Fuchs and Lauter 2000; Meyer 1979).

Using magnetic resonance imaging (MRI), a German scientist has undertaken an experiment to study the reactions provoked by imaginations of death. Seventeen young men were confronted with thoughts about death. As expected, the cerebral centers connected with fear (e.g., the right amygdala) showed increased activity, a sign that death confronts everyone ("does not matter to me" is a protection allegation). However, surprisingly, there was increased activity in the caudate nucleus also. The caudate nucleus, part of basal ganglia, acts as a switching center to connect the long nervous tracts between the peripheral nervous system and cerebral cortex, forming automated movement- and behavior patterns. It is speculated that the caudate nucleus helps oneself to act in conformity with the conventions of one's cultural surroundings. Philosopher Martin Heidegger already had postulated that such conformity may compensate for trepidation caused by fear of death.

A terror theory was proposed based on such behavior patterns. According to the terror theory, when confronted with a foreign culture, the individual feels his/her refuge is threatened and strikes back with all means at hand. Thus, the root of terrorism would not be bravery but fear of death (see at Liechty 2002; Quirin 2011; Santaniello 2011; see at Solomon et al. 1991).

In light of such developments, it could be stated that cryonics plays an important role in presenting the alternative to death. We are to ask teachers how one can introduce cryonics into school teaching. One of them, Jan Welke, wrote an interesting article about it in our first symposium volume (2013) in this series. The difficulty, he states, consists in the fact that teachers, school authorities, and committees themselves are convinced of the inescapability of death.

Cryonicists should elaborate cryonics teaching materials themselves and offer it to the school authorities or publishers of schoolbooks. Reactions to cryonics may be widely influenced by one's understanding of coping with death. For example, the individual avoids being remembered of his/her own death. He/she may be afraid to be confronted with death by an empty promise. The imagination of being packed into ice makes him/her freeze. Some people do not want the concept of their whole life from birth to death, planned for around 80 years (earning degrees, founding a family, building a house, planning their burial, financing all of it, etc.), to be touched, especially since they have no alternative plans for a following life span.

The concept of life and its every day realization seems to be more important compared with an extension of the life span later, which furthermore seems totally unrealistic. In a German TV talk show, after a fair discussion concerning cryonics, the moderator remarked that one could now return to realistic work (Maischberger 2010). A number of people are afraid of damaging the insurance systems by extending the life span. In this context, we often hear death to be an essential part of life. How is that? It's the negation of life!

For other people, despite their well-being, life seems to be a negative experience, and they agree into the limitation of life span. In the end they want to come to rest. This may also explain the remark of the Swiss author Erwin Koch (2010 translated): "thank god for being mortal."

Devaluation of life here seems to allow for the acceptance of death. Human imagination is able to trust in God, allowing human beings a life after death. On the other hand, reanimation of a cryopreserved body by human action seems hard to imagine. This may also be due to lack of a supporting social consensus.

At least it seems to be odd if medics state cryopreservation, performed after "death," to be nonsense. This indicates insufficient study of the medical definitions of death. Possibly here, unconsciously, the medieval belief in an irreplaceable essence of life -which is definitely lost at death- is at work. This may be a power of vitality up to the midst of the twentieth century named "entelechia," a term, still used by the gerontology pioneer Max Bürger (Bürger 1960).

If one sees the spectacular stop of heart action by cooling and the even more spectacular revival of a human heart during heart surgery, as well as the revival of mind and soul (the emotional self), then one's belief in the irreversibility of organ failure (death) may be challenged.

Cryonics promises no paradise. It is the extension of the life span by medical means. It is the preservation of a human body following failure of its life-maintaining organs and processes. Cryonics proponents of course can believe in religion and metaphysics, without questioning them.

We may fail to convince people of the value of cryonics if we do not reform school teaching in the youngest years as soon as possible.

Biological dying

In the preceding paragraphs we have used a definition of death that meant definite extinction of the conscious individual existence or at least a state where this is unavoidable. In medical practice, however, we need to adopt a definition that connotes allowing pronouncement of death soon after organ failure. Such practical definitions may be no real descriptions of definite death. Rather, they represent criteria for certain steps of dying or loss of viability. They are based on the loss of organ function, inability of reanimation, and signs of decay. Today, we know that heart arrest may not mean death, since a number of patients recovers from it. If heart arrest is associated with failure of reanimation and silence in the electroencephalogram (EEG), death can be pronounced after more than 5–9 min. of ischemia.

In Germany, regulations governing death allow pronouncement, if there is no EEG available, only following observation of distinct signs of death like "postmortem" lividity and "postmortem" rigidity. At room temperature, lividity develops not earlier than 20 min. of ischemia.

If there is a partial survival of brain areas, then identifying the regions of the brain that need to function to regain consciousness is crucial. Without a chance to return to consciousness, a person is definitely dead. Still a matter of question and study is the state reached at which the brain does not allow to restore consciousness.

When is the brain really dead?

The human brain cannot be reanimated following more than 5–9 min. of ischemia. The question arises if cryonics suspension makes sense at all 15 min. later when lividity is observed first. Furthermore, does it make sense at all to preserve a human body 2 hours after announcement of death (in Germany physicians inquest may need this time)?

There exists evidence indicating that this limit can be overcome. It seems that a 9- min. limit of brain recovery is not found in all species. Animals survived for somewhat longer times without blood circulation at room temperature (Safar 1993).

There also exists no evidence that all nerve cells are dead following 5–9 minutes of oxygen deprivation. Safar (1993) found most neurons to be alive following 22 min. of ischemia; a small number had been dead. Failure of brain reanimation, however, does not depend on cell death exclusively.

In general, the most damaging effect of ischemia within the first or second hour is that cerebral blood flow is limited or constrained (Dawson et al. 1997; Hossmann and Zimmermann 1974; Pearson et al. 1977). In transplant organs the length of the ischemic period plays a crucial role for the outcome of transplantation (Opelz 1998). In the ischemic state, lactic acidosis causes endothelial cells to swell (Paljärvi et al. 1983). Increased vascular resistance, withstanding reanimation, thus, plays an important role.

On the other hand, resistance vessels are widened. Pharmaca engendered vasoconstriction (epinephrine), raised blood pressure, heparin, insulin, and antacids application led to reanimation even beyond the 9- min. limit (Brown and Boruteite 2002; Hossmann 1988; Ratych et al. 1987; Safar et al. 1976; Schaffner et al. 1999; Wu et al. 1998).

Furthermore, there exist reperfusion problems (De Groot and Rauen 2007; Hayashida et al. 2007; Safar 1993; Solenski et al. 2002). Dogs show substantial brain damage following 10 min. of circulation arrest and reperfusion (Radovsky et al. 1995). Restarting blood flow after more than about 10 min of ischemia is typically more damaging than the ischemia itself. Ischemia sets the stage for oxygen (when available by reperfusion) to generate free radicals and reactive oxygen species rather than to contribute to cellular energy production (Zweier and Talukder 2006). In addition to oxygen-generated free radicals, cytokines can be a significant source of reperfusion injury.

In "dead" monkeys, brains subjected to an hour of warm ischemia show short-term recovery if there is no reperfusion injury (Hossman et al. 1986). During ischemia blood cells stiffen and agglutinate. Swelling of endothelium by transformation of oxynitrite into peroxynitrite during reperfusion as well as leukocyte adhesion narrow the lumen. Endothelial cells become more severely damaged than nerve cells themselves this way.

Two to six hours of ischemia followed by 24 hours of reperfusion more than triple times increases infarct volume (Aronowski et al. 1997). In the course of reperfusion, problems with obliteration of blood vessels are more important inhibitors of reanimation than is the death of brain cells (Best 2008).

The longer the ischemia, the worse is the reperfusion injury to blood vessels because of free radicals and hemorrhage and the greater the chance of "no reflow" (impeded circulation). In the absence of circulation cardiopulmonary support or perfusion fails to start reanimation (out of B. Best (online): "Ischemia and Reperfusion Injury in Cryonics").

Taken together, this means that failure of reanimation, after a longer time following the stop of circulation, is not exclusively dependent on irreversible death of brain cells. It is also a consequence of changes that can be influenced.

Cellular survival

Stem cells

It is well known that stem cells of mammalian and human "postmortem" connective tissues can be propagated long times after "death", reaching around 10–30 passages or more in cell cultures (e.g., Sames 1980).

Neural stem cells also have been isolated "postmortem" in mammalian and human brains. Living and proliferating stem cells can be isolated from rat forebrains after heart arrest up to 6 days, especially if cooled. Neural stem cells able to differentiate can also be harvested from human "postmortem" brain tissue, spinal cord, and retina up to 1 day "postmortem" and show proliferation and divisions up to 70 passages in culture (review: Klassen et al. 2004; Lovell et al. 2006; Mansilla et al. 2013; Palmer et al. 2001; Schwartz et al. 2003; Verwer et al. 2002b; Xu et al. 2003). Stem cells are able to improve neuronal survival in cultured "postmortem" brain tissue (Wu et al. 2008). It is not clear if one can use them to repair vitrified "postmortal" tissue.

Human muscle stem cells from donors up to 90 years of age maintain regenerative activity as long as 17 days after general organ failure (Latil et al. 2012).

Inner ear tissues (including sensory epithelia) of neonatal mice show regular tissue structure and cellular morphology up to 10 days "postmortem". There are no substantial changes in the number of cells. After 10 days the number of dead cells reaches 1/3 to 1/2 of the number of living cells. Stem cells can be isolated (Senn et al. 2007).

In addition, neurons can be formed by postnatal astroglia as in the mouse postnatal cortex (Heinrich et al. 2011).

The last legion of cellular defence

It has been speculated that hypoxia, acidosis, lack of nutrients, and other "postmortem" shortages may, in contrast to other cells, stimulate stem cells or select them, thus, enriching these cells, and making them more robust and efficient compared to those in living tissues.

Forming such stem cells may represent an effort to repair tissues in the face of general damage. Following failure of repair, they become deeply dormant to maintain their abilities (see Mansilla et al. 2013).

Neurons

Survival of animal neurons following ischemia

As we know today, the overwhelming number of brain neurons die slowly. Neurons in rat brain slices recover by oxygen supply after 10 min of ischemia (Pichugin et al. 2006). There exists significant synaptic activity following some hours of heart arrest, for example, in slices of the rat taken from hippocampus 3 hours "postmortem" and reperfused. Those neurons showed significant spiking (Charpak and Audinat 1998; Leonard et al. 1991).

Adult rats subjected to cerebral ischemia show no signs of neuron death for 2 hours, and only by 6 hours more than 15% of neurons appear to be dead (Garcia et al. 1995). "Postmortem" mouse brains, subjected to 6 hours of room temperature and another 18 hours at 4°C, show half the neurons to be morphologically intact (Scheuerle et al. 1993).

Following arterial occlusion substantial amounts of dead neurons (15% of the population) are not found before 5–6 hours in the ischemic area; only single cells have been TUNEL-positive apoptotic ones. In the neocortex and hippocampus, the number of damaged cells increases significantly after 24 hours following the occlusion. Even at 72–96 hours following occlusion, single intact neurons have been found in the necrotic cortex area (Garcia et al. 1995; Pulsinelli et al. 1982; Rupalla et al. 1998).

Neuron-like cells were explanted from the frontal cortex of adult rats. They could be cultured when explanted following 6 hours "postmortem" at 22°C or even following 24 hours at 4°C, with 20% recovery, after each of the ischemic periods respectively. The tissue contained 160, 000 neurons per mg. Of the cells harvested immediately following sacrifice, 9% of the neurons originally present in the brain were viable. After 5 days in culture, 23–42% of the originally isolated neurons have been viable. Surviving neuron-like cells represented 0.5–2.75% of the cells contained in the brain (Viel et al. 2001). Lipton (1999) reviewed mechanisms of ischemic cell death in brain neurons.

Survival of ischemia by human neurons

The impossibility to reanimate the human brain following 5–9 min. of ischemia could be explained partially by the fact that cells enter apoptosis. Neurons of human brain tissue extracted 3–6 hours "postmortem" have been shown to recover oxidative metabolism and axon transport after suitable in-vitro treatment (Dai et al. 1998).

Out of human brain tissue from the parietal cortex, taken 2.6 hours "postmortem" on an average, neurons have been isolated. More than 82% of the isolated cells were living. However, only the fraction richest in neurons has been evaluated, and there was a lot of debris suspicious to contain disintegrated cellular material (Konishi et al. 2002).

Slices of human "postmortem" brains taken up to 8 hours "postmortem" contain living cells and can be used to prepare tissue cultures even after global organ failure. In cultured cortical brain slices taken 2–8 hours "postmortem" (average 4.2 hours), including periods at room temperature (1–2 hours during transport), neurons could be kept alive. Of the cells, 30–50% (without separation of different cell types) were alive; 20–25% were damaged but not dead.

One of the most exciting findings is that dead cells seemingly remain with minor changes in their location, a fact that could be favorable for tissue reconstruction (Verwer et al. 2002a;, 2003). In the cultures the cells lived for weeks and can thus be used for experiments.

Survival of cells in brain slices

State of cells	Patients		
	1	2	3
viable	8,75	10,00	8,75
leaky	7,50	8,75	6,25
viable + leaky	16,25	18,75	15,00
dead	25,00	20,00	10,00
total	41,25	38,75	25,00

Table legend: rough mean values taken from Verwer et al. (2002a)

Cellular damage and death following ischemia

There exist two main types of cellular death: necrosis and apoptosis

Necrosis can lead to rapid dying, consumes no energy, is a passive event, and is not DNA -governed. The main damage is to the cell membrane. *Apoptosis* is the genetically programmed suicide of cells. As a rule it is an active process needing energy and time (see figure below). The main damage is to the chromosome. Excitotoxicity seems to be responsible for necrosis, while apoptosis is a separate process (Choi 1996).

Rat neurons at 9 hours after global organ failure at room temperature showed first symptoms of autolysis. Ribosomes disappeared and neurons strongly staining for the pro-apoptotic enzyme caspase-3 reached 2.5% (Sheleg et al. 2008).

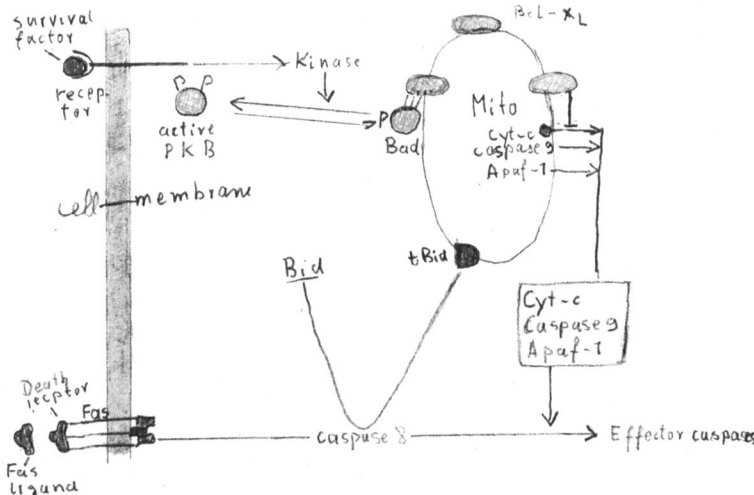

Fig. Legend: Apoptosis, simplified diagram adapted from Fig. 20–6 in Lehrbuch Vorklinik part A, Deutscher Ärzteverlag, Köln 2003.
Caspases = cysteine- dependent aspartate specific proteins, Cyt-c = Cytochrome C, Apaf-1 = apoptotic protease activating factor 1, Bid = Bh3-interacting domain, Bcl-XL (Bcl = antiapoptotic B-cell-lymphoma factors), Bad = antiapoptotic factor, Mito = mitochondrium (Figure hand drawn by the author)

Three ways of apoptosis stimulation may be active: one via Fas, one via the mitochondrium (intrinsic or mitochondrial pathway), and—not shown in the figure—the other via direct transmembrane stimuli. Fas ligands such as cytokines, for example, tumor necrosis factor (TNF-alpha), lead to activation of initiator caspases like caspase-8, which in turn activates effector caspases such as caspase-3. The mitochondrial pathway (via a mitochondrial triple complex of Cyt-c, caspase-9, and Apaf-1 named apoptosome) activates caspase -3. Cyt-c goes into the cytosol where the apoptosome is formed (square framed in the figure). Its release is a function of tBid (by heterodimerizing with Bcl). Bid is activated by caspase -8 to tBid. Bcl inhibits Cyt-c release.

Fourteen or more caspases are known. Pro- and antiapoptotic factors form many rebound loops to regulate cell death. For example, Bax, Bad, and Bid are proapoptotic; Bcl2, Bcl-Xl, and Ced 9 are antiapoptotic.

External apoptose-inhibiting factors (survival factors) such as growth factors act via a specific membrane receptor activating a kinase, which forms (unphosphorilated) Bad interacting with Bcl-XL. This inhibits the action of the apoptosome and therewith the activation of the effector caspases. Effector caspases are destructive.

Delayed cell death following ischemia is an active process. It reveals some signs of apoptosis. Bax is expressed before delayed death of cells (Chen et al. 1996).

In human brain infarction, apoptosis begins (as revealed by increase of caspase-3) in the first 2 days following infarction, but is not totally completed. The morphology of the dead cells is rather similar to necrosis and some DNA splitting as a symptom of apoptosis develops late (Love et al. 2000).

In other organs and cell types a retarded start of apoptosis or incomplete apoptosis has also been discovered, following different forms of ischemia. In excised human kidneys, during 85 min. of ischemia at 37°C, the substances Bax and caspase-9, which react with mitochondria in favor of apoptosis, increased. Proteins with inhibiting action on apoptosis, such as Bcl2 or cFLIP, decreased. However, the way to the most destructive caspase-3 was not activated (even not via the so-called death receptor). Apoptosis remained incomplete (Wolfs et al. 2005). Some different variations of apoptosis have been described as more or less active processes with and without the involvement of caspase-3 (Yakovlev and Faden 2004).

How can we explain this cell preservation? At the start of ischemia, oxygen is exhausted and the reserves of energy, namely activated phosphates like ATP as well as sugars, are also totally consumed after 2 min. There is no energy left to complete apoptosis, and macrophages and other cells that normally ingest and degrade dead cells are also exhausted of energy. However, as long as a high number of brain neurons can recover, apoptosis cannot have proceeded very far some hours after global organ failure.

The chronological pattern of apoptosis differs between different tissues.

In photo receptors of the rat, light-mediated apoptosis has been found following 90 min. of ischemia. Thereby, DNA breaks follow the morphological changes. In the neighboring pigmented epithelium of the eye, signs of apoptosis were seen many hours later (Hafezi et al. 1997). Thus, during total ischemia apoptosis takes place in different ways and its procedure seems to remain incomplete.

These findings are still very sparse, especially as far as quantification of cell numbers is concerned. In case they become corroborated, one may hope that brain reconstruction is possible in principle. As mentioned above, dead cells remain in their proper location. The brain throughout shows a nearly unchanged structure in the light microscope.

We always had known that the "postmortem" organs are well preserved if studied at light microscopic magnifications. We teach our students using slices of human organs partially taken some times after death, and perform pathological diagnoses on such materials.

A majority of the cryonics methods are taken from clinical procedures, especially from those of heart surgery, including reanimation of patients by heart massage and oxygen supply. However, cryonics differs from heart surgery. We now give no oxygen to cryonics patients, since we do not intend total recovery of life before vitrification. Thus, we do not need to tolerate oxygen- engendered damage. In experimental suspended animation, deprivation of oxygen has been shown to lead to a dormant state of cells. Recently it has been shown that following general organ failure muscle stem cells too can fall into a dormant state (after maintenance of regenerative activity as long as 17 days of ischemia and obvious exhaustion; see above and Latil et al. 2012).

Cooling

Since organ activity no longer produces heat, in the first stages of ischemia spontaneous fast cooling takes place. Newton's law of cooling dictates that temperature drop is most rapid upon initial application of cooling. So there is a natural drop in brain temperature associated with the stop of blood flow. It is commonly noted that—as a rule—metabolic rate is halved for every 10°C drop in temperature. But due to reduction of lipid peroxidation, reduction of temperature has a protective effect exceeding reduction of metabolism. Experiments on gerbils indicate that a drop in temperature from 37 to 31°C nearly triples the amount of time of of ischemia that neurons can tolerate (Takeda et al. 2003). Dogs cooled to 20°C can withstand 60 min of ischemia and at 10°C 120 min of ischemia (Behringer et al. 2003).

Temperatures below 15°C considerably reduced ischemic oxidative stress in mice (Khandoga et al. 2003). A temperature reduction from 37 to 26°C completely inhibited potassium-induced neurotransmitter release from rat astrocytes (Kimelberg et al. 1995). Marked increases in nitric oxide end-products, caused by glutamate infusion in rats, were completely eliminated by reducing the temperature from 37 to 32°C (Shima et al. 2003).

Rats reperfused after a 15-min ischemic period had over three times as many hydroxyl radicals 1 hour later than rats subjected to ischemia without reperfusion. But rats reperfused at 30°C, rather than 36°C, had half as many hydroxyl radicals as rats perfused at 36°C (Ki et al. 1996).

The protective effects of hypothermia against ischemic damage are non linear with change in temperature. Thereby and by further cooling, enzymes too, are inhibited. Thus, autolysis is retarded.

Cold ischemia

Nonetheless, more than a day or two of cold ischemia (4°C) greatly reduces survival of kidneys held in organ preservation solution. In rats, following unpublished results of Y. Pichugin, the viability of cultured cells in hippocampal brain slices has been tested using the potassium:sodium relation. The test does not discern neurons from other cell types. It is also not clear if all cells show the same slow changes or if part of the population is dying. Decrease in viability to the extent of 45–47% is exhibited during warm ischemia (room temperature) (according to literature) after 3 hours.

In contrast, during cold ischemia (2–4°C), after 4 hours 80–90% of cells had survived, viability decreasing slowly to 50% at around 10 hours and to 20% at more than 50 hours. If 3 hours of warm ischemia are followed by 21–24 hours of cold ischemia (2–4°C), survival is reduced to 20%. Cold ischemia—by increasing chelatable iron, which opens mitochondrial transition pores—favors necrosis and apoptosis (Rauen et al. 2004).

We can conclude that up to at least 8 hours after global organ failure, a high number of neurons (between 20 and 50% or more) are alive and able to produce electrical signals. The dead cells remain in their original situation. Damage in the dead cells and tissue components can be shown only by electron microscopy.

However, the overall structure is preserved. Neuronal stem cells remaining able to differentiate outlive ischemia longer than the neurons themselves One can speculate if such a brain, given a restored blood circulation, would be able to repair itself if apoptosis could be stopped.

We know that an aged brain does not repair age-related damage spontaneously even if it seems to possess the means to do so. Thus, a damaged brain also needs medical help to be repaired by its own stem cells or, for example, by nanotechnology. Even if there are cell losses and partial decomposition of structures, a reconstruction could become possible in the future. The results of analysis of whole brain connectomics as a model of the general arrangement of cells and structures may allow such reconstruction (McIntyre and Fahy 2015; Mikula 2016). If the losses of structure and cells are small, even individuality may be restored. Stem cells may be able to replace not only neurons but also the functions of those neurons. These new neurons may be able to reconstruct original structure and function of the individual brain by exchange of information with neurons remaining alive. Astroglia cells are able to differentiate into neurons. However, we do not have enough information about the "postmortem" fate of such cells (Kriegstein and Alvarez-Buylla 2009). Embryonic stem cells are also able to differentiate into neurons forming neuronal networks (see, e.g., Bühnemann et al. 2006).

Astonishingly, cells differentiated this way are able to appropriately project axons to subcortical targets (Ideguchi et al. 2010).

Now a new definition of death comes into sight—a negative one. As long as we can expect that repair is possible, the brain is not dead.

Chances of intervention

There are already means to influence apoptosis, but targeted reconstitution of apoptotic cells is only visible in first rates, for example caspase-3-inhibitor Z-DEVD-FMK favors survival of neurons in the ischemic rat hippocampus (Chen et al. 1998)

Blockade of caspase does stop apoptosis temporarily but not totally (Volbracht et al. 2001). In experimental models of stroke, caspase inhibition affords protection in certain neuronal populations but not in others (Zhan et al. 2001).

Another chance to favor survival of neurons in human cortical brain tissues—dissected after up to 9.5 hours "postmortem". (including 1–2 hours of warm ischemia during transport)—has been co-cultivation with embryonic stem cells of rats. The donors concerned have been aged up to 94 years. The number of dead cells in controls was 17 per 26/cmm,; some more than 3/cmm have been vital, while the others showed damage. Of the co-cultivated cells, 6 of 16 neurons have been vital (Wu et al. 2008).

Pharmacological treatment may also favor survival. Tthe radical trapping agent NXY-059 in a model of transient ischemia in rats reduced infarct volume and decreased neurological impairment (Sydserff et al. 2002). Insulin activates the PI3K-Akt survival pathway in vulnerable neurons following global brain ischemia (Sanderson et al. 2009).

The substantial capacity for restitution of function in the living brain, even following abundant cell losses, let us hope for a possible reconstruction of the brain following vitrification. Redundant contents of memory and high neuron numbers—but in each case so-called plasticity of brain cells—play a role (Carmichael 2006; Chen et al. 1998; Dancause et al. 2005; Nudo 2007; Wu et al. 2008). Normally, cells completing apoptosis are phagocytosed. However, the "postmortem" appearance of the brain cells is not generally like those that have completed apoptosis and/or that have been phagocytosed.

We need more studies, especially counts of viable cells with different viability criteria, to exclude error by use of different methods. We are still unable to define memory as well as consciousness and what has to be preserved to maintain individuality.

Furthermore, since we do not know the facilities of reconstruction available for future development, it makes sense to be generous with "postmortem" time before suspension.

Survival of subzero cooling

A complete neurological recovery of hamsters, 60% of whose brain water had been ice, was found by Audrey and Smith (Lovelock and Smith 1956).

Thus, the brain is an especially ice-crystal- tolerant organ insofar as 60% water as ice-crystals causes little gross damage. Even without cryoprotectants, 80% of synapses in whole brain tissues cooled to $-70°C$ retain the metabolic properties of fresh brain biopsy synapses (Hardy et al. 1983).

Experiments with cats indicate that cat brains cooled to $-20°C$ in 15% v/v glycerol (62% brain water as ice) for 777 days and 7.25 years, show normal-looking EEG patterns upon rewarming. However, neurological activity is less for the 7.25-year brains (Suda et al. 1966,; 1974). The results (of the possibly cruel experiments) have not been reproduced up to now.

Following Ben Best: vitrification in cryonics (online) hemorrhage and cell loss in these brains, which were found at least following the 7.25-year storage, probably could have been prevented using several measures: (1) addition of glucose (nutrient) to the perfusion fluid, (2) careful washing of glycerol from the brains as part of the thawing/reperfusion process, and (3) storing the brains at lower temperatures with higher glycerol concentrations.

Conclusion

Following our definition, death is not described by loss or maintenance of any part of our body. Otherwise, one could ask if a human being is dead as long as we can reconstitute his DNA from a bone fragment. A human being is dead if there is no chance to reinstall self-consciousness and memory. Self-identity is maintained by the DNA in connection with memory as well as individual structural and psychic particularities. For people dying today perfect cooling centered to the brain is essential. Development of conditions for a successful revitalization may need centuries of time. Cryonics provides this time.

References

Aronowski J, Strong R, Grotta JC (1997) Reperfusion injury: demonstration of brain damage produced by reperfusion after transient focal ischemia in rats. J. Cereb Blood F Met 17: 1048–56 doi:10.1097/00004647-199710000-00006

Behringer W, Safar P, Wu X et al. (2003) Survival without brain damage after clinical death of 60-120 minutes in dogs using suspended animation by profound hypothermia. Crit Care Med 31: 1523–31

Best BP (2008) Scientific justification of cryonics practice. Rejuv Res 11: 493-503

Brown GC, Borutaite V (2002) Nitric oxide inhibition of mitochondrial respiration and its role in cell death. Free Rad Biol Med 33: 1440–50

Bürger M (1960) Altern und Krankheit als Problem der Biomorphose. Georg Thieme Verlag, Leipzig

Bürgin D (1981) Das Kind, die lebensbedrohende Krankheit und der Tod. Huber, Bern

Bühnemann C, Scholz A, Bernreuther C et al. (2006) Neuronal differentiation of transplanted embryonic stem cell-derived precursor in: Stroke lesions of adult rats. Brain 129: 3238–48

Carmichael ST (2006) Cellular and molecular mechanisms of neural repair after stroke: making waves. Ann Neurol 59: 735–42

Charpak S, Audinat E (1998) Cardiac arrest in rodents: maximal duration compatible with a recovery of neuronal activity. Pub Nat Acad Sci U S A 95: 4748–53

Chen J, Nagayama T, Jin K et al. (1998) Induction of caspase-3-like protease may mediate delayed neuronal death in the hippocampus after transient cerebral ischemia. J Neurosci 18: 4914–28

Chen J, Zhu RL, Nakayama M et al. (1996) Expression of the apoptosis-effector gene Bax, is up-regulated in vulnerable hippocampal CA1. J Neurochem 67: 64–71

Choi DW (1996) Ischemia induced neuronal apoptosis. Curr Opin Neurobiol 6: 667–72

Dai JP, Swaab DF, Buijs RM (1998) Recovery of axonal transport in "dead neurons". Lancet 351: 499–500

Dancause N, Barbay S, Frost SB et al. (2005) Extensive cortical rewiring after brain injury. J. Neurosci. 25: 10167–79

Dawson DA, Ruetzler CA, Hallenbeck JM (1997) Temporal impairment of microcirculatory perfusion following focal cerebral ischemia in the spontaneously hypertensive rat. Brain Res 749: 200–8

De Groot H, Rauen U (2007) Ischemia-reperfusion injury: processes in pathogenetic networks: a review. Transpl Proc 39: 481–4

Fuchs T, Lauter H (2000) Psychiatrische Aspekte des Lebensendes. In: Helmchen H, Henn F, Lauter H et al. (eds.) Psychiatrie der Gegenwart, vol 6, 4rd edn, Springer, Berlin Heidelberg New York 291–309

Garcia JH, Liu KF, Ho KL (1995) Neuronal necrosis after middle cerebral artery occlusion in Wistar rats progresses at different time intervals in the caudoputamen and the cortex. Stroke 26: 636–42

Hafezi F, Marti A, Munz K et al. (1997) Light-induced apoptosis: differential timing in the retina and pigment epithelium. Exp Eye Res 64: 963–70

Hardy JA, Dodd PR, Oakley AE et al. (1983) Metabolically active synaptosomes can be prepared from frozen rat and human brain, J Neurochem 40: 608–14.
DOI: 10.1111/j.1471-4159.1983.tb08024.x

Hayashida M, Sekyiama H, Orii R et al. (2007) Effects of deep hypothermic circulatory arrest with retrograde cerebral perfusion on electroencephalographic bispectral index and suppression ratio. J Cardiothorac Vasc Anesth 21: 61–7

Heinrich C, Gascón S, Masserdotti G et al. (2011) Generation of subtype specific neurons from postnatal astroglia of the mouse cerebral cortex. Nature Protoc 6: 214–28

Hossmann KA (1988) Resuscitation potentials after prolonged global cerebral ischemia in cats. Crit Care Med 16: 964–71

Hossmann KA, Grosse-Ophoff B (1986) Recovery of monkey brain after prolonged ischemia. I. Electrophysiology and brain electrolytes. J Cereb Blood F Met 6: 15–21

Hossmann KA, Zimmermann V (1974) Resuscitation of the monkey brain after 1h complete ischemia. I. Physiological and morphological observations. Brain Res 81: 59–74

Ideguchi M, Palmer TD, Recht LD et al. (2010) Murine embryonic stem cell-derived pyramidal neurons integrate into the cerebral cortex and appropriately project axons to subcortical targets. J Neurosci 30: 894–904

Khandoga A, Enders G, Luchting B et al. (2003) Impact of intraischemic temperature on oxidative stress during hepatic reperfusion. Free Rad Biol Med 35: 901–9
DOI: 10.1016/S0891-5849(03)00430-1

Ki HY, Zhang J, Pantadosi CA (1996) brain temperature alters hydroxyl radical production during cerebral ischemia/reperfusion in rats. J Cereb Blood F Met 16: 100–6

Kimelberg HK, Rutledge E, Goderie S et al. (1995) Astrocytes swelling due to hypotonic or high K^+ medium causes inhibition of glutamate and aspartate uptake and increases their release. J Cereb Blood F Met 15: 409–16
doi:10.1038/jcbfm.1995.51

Klassen H, Boback Z, Kirov II et al (2004) Isolation of retinal progenitor cells from post-mortem human tissue and comparison with autologous brain progenitors. J Neurosci Res 77: 334–43 (/doi/10.1002/jnr.v77:3/issuetoc)

Koch E (2010) Die Eismenschen. GEO (04) 2010: 18–117

Konishi Y, Lindholm K, Yang LB et al. (2002) Isolation of living neurons from humane elderly brains using the immunomagnetic sorting DNA-linker system. Amer J Path 161: 1567–76

Kriegstein A, Alvarez-Buylla A (2009) The glial nature of embryonic and adult neural stem cells. Ann Rev Neurosci 32: 149–84

Latil M, Rocheteau P Chatre L et al (2012) Skeletal muscle stem cells adopt a dormant cell state post mortem and retain regenerative capacity. Nat Commun 3: 903 doi:10.1038/ncomms1890

Leonard BW, Barnes CA, Rao G et al. (1991) The influence of postmortem delay on evoked hippocampal field potentials in the in vitro slice preparation. Exper Neurol 113: 373–7

Liechty D (ed.) (2002) Death and denial. Interdisciplinary perspectives on the legacy of Ernest Becker. D.1st edn. Praeger, Westport CT

Lipton P (1999) Ischemic cell death in brain neurons. Physiol Rev 79: 1431–568

Love S, Barber R, Wilcock GK (2000) Neuronal death in brain infarcts in man. Neuropathol Appl Neurobiol 26: 55–66

Lovell MA, Geiger H, Van Zant GE et al. (2006) Isolation of neural precursor cells from Alzheimer's disease and aged control postmortem brain. Neurobiol Aging 27: 909–17

Lovelock JE, Smith AU (1956) Studies on golden hamsters during cooling to and rewarming from body temperatures below 0 degrees C. III. Biophysical aspects and general discussion. Proc Royal Soc B 145: 427–42

Mansilla E, Mártire K, Roque G et al. (2013) Salvage of cadaver stem cells (CSCs) as a routine procedure: history or future for regenerative medicine. J Transplant Technol Res 3: 118 doi.org/10.4172/2161-0991.1000118

McIntyre RL, Fahy GM; Aldehyde stabilized cryopreservation, in this volume

Meyer JE (1979) Todesangst und das Todesbewußtsein der Gegenwart. Springer, Berlin Heidelberg New York.

Mikula S (2016) Progress towards mammalian whole-brain cellular connectomics. Front Neuroanat 10: 62 Published online 2016 Jun 30.
doi:10.3389/fnana.2016.00062

Nudo RJ (2007) Postinfarct cortical plasticity and behavioral recovery. Stroke 38: iss 2 Suppl 840–5

Opelz G (1998) For the Collaborative Transplant Study (CTS), Cadaver kidney graft outcome in relation to ischemia time and HLA match. Transplant Proc 30: 4294–6

Paljärvi L, Rehncrona S, Soderfeldt B et al. (1983) Brain lactic acidosis and ischemic cell damage: quantitative ultrastructural changes in capillaries of rat cerebral cortex. Acta Neuropathol 60: 232–40

Palmer TD, Schwartz PH, Taupin PH et al. (2001) Cell culture: progenitor cells from human brain after death. Nature 411: 42–3

Pearson J, Korein J, Harris JM et al. (1977) Brain death: II. Neuropathological correlation with the radioisotopic bolus technique for evaluation of critical deficit of cerebral blood flow. Ann Neurol 2: 206–10

Pichugin Y, Fahy GM, Morin R. (2006) Cryopreservation of rat hippocampal slices by vitrification. Cryobiology 52: 228–40

Pulsinelli WA, Brierley JM, Plum F (1982) Temporal profile of neuronal damage. Ann Neurol 11: 491–8

Quirin MR, University Osnabrück, press release Nr. 90/2011 of march 04 2011

Radovski A, Safar P, Sterz F et al. (1995) Regional prevalence and distribution of ischemic neurons in dog brains 96 hours after cardiac arrest of 0 to 20 minutes. Stroke 26: 2127–33

Rauen U, Petrat F, Sustmann R, De Groot H (2004) Iron-induced mitochondrial permeability transition in cultured hepatocytes. J Hepatol 40: 607–15

Ratych RE, Chuknyiska RS, Bulkley GB (1987) The primary localization of free radical generation after anoxia/reoxygenation in isolated endothelial cells. Surgery 102: 122–31

Rupalla K, Allegrini PR, Sauer D, Wiessner C (1998) Time course of microglia activation and apoptosis in various brain regions after permanent focal cerebral ischemia in mice. Acta Neuropathol 96: 172–8

Safar P, Stezoski W, Nemoto EM (1976) Amelioration of brain damage after 12 minutes cardiac arrest in dogs. Arch Neurol 33: 91–5

Sames K (1980) Morphologische und histochemische Untersuchungen über das in-vitro- und in-vivo Altern von Corneaendothel und Trabeculum corneosclerale. Habilitation, Erlangen

Sanderson TH, Kumar R, Murariu-Dobrin AC et al. (2009) Insulin activates the PI3K-Akt survival pathway in vulnerable neurons following global brain ischemia. Neurolog Res 31: 947–58

Santaniello K (2011) Terror Management und Affekt- psychophysiiologische Prozesse und individuelles Affektmanagement bei Mortalitätssalienz. Osnabrück (Dissertation)

Schaffner DH, Eleff SM, Brambrin AM et al. (1999) Effect of arrest time and cerebral perfusion pressure during cardiopulmonary resuscitation on cerebral blood flow, metabolism, adenosine triphosphate recovery, and pH in dogs. Crit Care Med 27: 1335–42

Scheuerle A, Pavenstaedt I, Schlenk R et al. (1993) In situ autolysis of mouse brain: ultrastructure of mitochondria and the function of oxidative phosphorylation and mitochondrial DNA. Virchows Arch B 63: 331–4

Schwartz PH, Bryant PJ, Fuja TJ et al. (2003) Isolation and characterization of neural progenitor cells from post-mortem human cortex. J Neurosci Res 74: 838–51

Senn P, Oshima K, Teo D et al (2007) Robust postmortem survival of murine vestibular and cochlear stem cells. JARO N: 191–204

Sheleg SV, LoBello JR, Hixon H et al. (2008) Stability and autolysis of cortical neurons in post-mortem adult rat brains. Int J Clin Exp Pathol 1: 291–9

Shima H, Fujisawa H, Suehiro E, et al. (2003) Mild hypothermia inhibits exogenous glutamate-induced increases in nitric oxide synthesis. J Neurotrauma 20: 1179–87

Solenski NJ, diPierro CG, Trimmer PA et al. (2002) Ultrastructural changes of neuronal mitochondria after transient and permanent cerebral ischemia. Stroke 33: 816–24

Solomon S, Greenberg J, Pyszczynski T (1991) A terror management theory of social behavior: The psychological functions of self-esteem and cultural worldviews. In: Zanna MP (ed.) Adv Exp Soc Psychol 24: 93–159

Suda I, Adachi C, Kito K (1966) Viability of long term frozen cat brain in vitro. Nature 212: 268–70
doi:10.1038/212268a0

Suda I, Kito K, Adachi C (1974) Bioelectric discharges of isolated cat brain after revival from years of frozen storage. Brain Res 70; 527–31

Sydserff SG, Borelli AR, Green AR et al. (2002) Effect of NXY-059 on infarct volume after transient or permanent middle cerebral artery occlusion in the rat; studies on dose, plasma concentration and therapeutic time window. Br J Pharmacol 135: 103–12

Takeda Y, Namba K, Higuchi T, et al. (2003) Quantitative evaluation of the neuroprotective effects of hypothermia ranging from 34°C to 31°C on brain ischemia in gerbils and determination of the mechanism of neuroprotection. Crit Care Med 31: 255–60

Verwer RWH, Baker RE, Boiten EFM et al. (2003) Post mortem brain tissue cultures from elderly control subjects and patients with a neurodegenerative disease. Exper Gerontol 38: 167–72

Verwer RWH, Dubelaar EJG, Hermens WTJMC et al. (2002) Tissue cultures from adult human postmortem subcortical brain areas. J Cell Mol Med 6: 429–32

Verwer RWH, Hermens WTJMC, Dijkhuizen PA et al. (2002a) Cells in human postmortem brain tissue slices remain alive for several weeks in culture. FASEB J 16: 54–60

Viel JJ, McManus DQ, Cady C. et al. (2001) Temperature and time interval for culture of postmortem neurons from adult rat cortex. J Neurosci Res 64: 311–21

Volbracht C, Leist M, Kolb SA Nicotera P (2001) Apoptosis in caspase-inhibited neurons. Mol Med 7: 36–48

Welke J (2013) Cryonics in school education. In: Sames KH (ed.) Applied human cryobiology, Vol 1 Applied cryobiology—human biostasis. Ibidem, Stuttgart 117–21

Wolfs TG, de Vries B, Walter SJ et al. (2005) Apoptotic cell death is initiated during normothermic ischemia in human kidneys. Am J Transplant 5: 68–75

Wu L, Sluiter AA, Guo HF et al. (2008) Neural stem cells improve neuronal survival in cultured postmortem brain tissue from aged and Alzheimer patients. J Cell Mol Med 12: 1611–21

Wu S, Tamaki N, Nagashima T et al. (1998) Reactive oxygen species in reoxygenation injury of rat brain capillary endothelial cells. Neurosurg 43: 577–83, discussion 584

Xu Y, Kimura K, Matsumoto N et al. (2003) Isolation of neural stem cells from the forebrain of deceased early postnatal and adult rats with protracted post-mortem intervals. J Neurosci Res 74: 533–40

Yakovlev AG, Faden I (2004) Mechanisms of Neural Cell Death: Implications for Development of Neuroprotective Treatment Strategies. NeuroRx 1: 5–16.
doi: 10.1602/neurorx.1.1.5

Zhan RZ, Wu CH, Fujihara H et al. (2001) Both Caspase-dependent and caspase-independent pathways may be involved in hippocampal CA1 neuronal death because of loss of cytochrome c from mitochondria in a rat forebrain ischemia model. J Cereb Blood F Met 21: 529–40
doi:10.1097/00004647-200105000-00007

Zweier JL, Talukder MAH (2006) The role of oxidants and free radicals in reperfusion injury. Cardiovasc Res 70: 181–90

Contributors

Artyuhov, Igor V
Institute of Biology of Aging, Moscow, 125284, Russia

Artyukhov VI
Department of Materials Science and NanoEngineering, Rice University, Houston, TX, USA

Barroso, Pablo
Escuela Superior de Ingeniera, University of Seville, C/Descubrimientos s/n 41092 Seville (Spain)

Best, Benjamin P. Bsc, BBA
Life extension foundation, Fort Lauderdale, Florida USA

Buslow, Dmitry
Institute of Biology of Aging, Moscow, 125284, Russia

De Wolf, Aschwin
CEO of Advanced Neural Biosciences, Inc., Portland, Oregon USA

Fahy, Gregory Ph.D.
CSO, 21st Century Medicine, 14960 Hilton Drive, Fontana, CA 92336, USA

Gouras, Peter MD, Professor of Ophthalmology
Dept of Ophthalmology, Columbia University New York

Mathwig, Klaus Ph.D. Assistant Professor for Analytical Chemistry
Groningen Research Institute of Pharmacy, University of Groningen, The Netherlands

McIntyre, Robert L.
Nectome, Inc., 2323 Market Street, Unit A, Oakland, CA 94607, USA

Nemitz, Dirk M.Sc.
University Duisburg-Essen

Olmo, Alberto
Department of Electronic Technology, University of Seville, Avda, Reina Mercedes s/n 41010 Seville (Spain)

Peregudov A.
Institute of Biology of Aging, Moscow, 125284, Russia. E-mail: pulver@bioaging.ru

Pulver, Alexander
Institute of Biology of Aging, Moscow, 125284, Russia.

Risco, Ramon Ph.D
National Accelerators Center, CNA-CSIC, C/Descubrimientos s/n 41092 Seville (Spain)

Sames, Klaus Hermann MD Professor (German APL, Anatomy, Gerontology)
D-89250 Senden Iller Germany

Saul, Nadine Ph.D
Humboldt-Universität zu Berlin, Institute of Biology, Molecular Genetics Group, Philippstr. 11-13, 10115 Berlin, Germany

Shamaev, NV
Institute of Biology of Aging, Moscow, 125284, Russia

Tselikovsky A.V
N.N. Burdenko Voronezh State Medical Academy, Voronezh, Russia

ibidem.eu